SUSTAINABLE SOILS
RE-ENGINEERING

KENNEDY C. ONYELOWE • JULIAN C. ARIRIGUZO • CHARLES N. EZUGWU

PARTRIDGE

Copyright © 2019 by KENNEDY C. ONYELOWE, JULIAN C. ARIRIGUZO, CHARLES N. EZUGWU

ISBN:	Softcover	978-1-5437-5099-7
	eBook	978-1-5437-5100-0

All rights reserved. No part of this book may be used or reproduced by any means, graphic, electronic, or mechanical, including photocopying, recording, taping or by any information storage retrieval system without the written permission of the author except in the case of brief quotations embodied in critical articles and reviews.

Because of the dynamic nature of the Internet, any web addresses or links contained in this book may have changed since publication and may no longer be valid. The views expressed in this work are solely those of the author and do not necessarily reflect the views of the publisher, and the publisher hereby disclaims any responsibility for them.

Print information available on the last page.

To order additional copies of this book, contact
Toll Free 800 101 2657 (Singapore)
Toll Free 1 800 81 7340 (Malaysia)
orders.singapore@partridgepublishing.com

www.partridgepublishing.com/singapore

SUSTAINABLE SOILS RE-ENGINEERING

KENNEDY C. ONYELOWE
B. Eng., M. Eng., Ph.D., R. Eng., M.IGS

of

Department of Civil Engineering, College of Engineering and Engineering Technology, Michael Okpara University of Agriculture, Umudike, Umuahia, Nigeria

Adjunct Senior Lecturer, Department of Civil Engineering, Faculty of Engineering and Technology, Alex Ekwueme Federal University, Ndufu-Alike, Ikwo, Abakiliki, Nigeria

Principal Researcher, Research Group of Geotechnical Engineering, Construction Materials and Sustainability, Hanoi University of Mining and Geology, Hanoi, Vietnam.

Lead Researcher, Research Group of Solid Waste and Biomass Recycling for Ecofriendly, Eco-Efficient and Sustainable Soil, Concrete and Pavement Re-Engineering, Michael Okpara University of Agriculture, Umudike, Nigeria.

JULIAN C. ARIRIGUZO
B. Eng., M. Eng., Ph.D., R. Eng.

of

Department of Mechanical Engineering, College of Engineering and Engineering Technology, Michael Okpara University of Agriculture, Umudike, Umuahia, Nigeria

CHARLES N. EZUGWU
B. Eng., M. Eng., Ph.D., R. Eng.

of

Department of Civil Engineering, Faculty of Engineering and Technology, Alex Ekwueme Federal University, Ndufu-Alike, Ikwo, Nigeria

CONTENTS

Preface xvii
Notation Index...xix

1. Background 1

2. Sustainability In Soils Re-Engineering 7

3. Sustainable Materials Processing 11
   3.1 Processing Ash by Direct Combustion 11
   3.2 Processing Powder by Direct Crushing...............................13
   3.3 Processing Coupled Materials by Blending........................13

4. Soils And Materials Blending 15
   4.1 Soils, Materials and Proportionate Mixing 15
   4.2 Diffused Double Layer Reactions and the Adsorbed Complex 17

5. Sustainable Soils Properties Improvement...................... 20
   5.1 Gradation and Characterization of Selected Soils 20
   5.1 Consistency..22
   5.2 Compaction 30
   5.3 Volumetric Changes...49
   5.4 Sorptivity 55
   5.5 Erodibility..62
   5.6 California Bearing Ratio 65

    5.7 Unconfined Compressive Strength and Durability 70

    5.8 Resistance Value 78

    5.9 Resilient Modulus 87

6. Modeling And Optimization Techniques In Soils Re-Engineering 90

    6.1 Analysis of Variance 90

    6.2 Multiple Regression and Nonlinear Multiple Regression 92

    6.3 Scheffe's Method for Sustainable Soils Re-Engineering 93

    6.5 Extreme Vertex Design for Sustainable Soils Re-Engineering 101

7. Environmental Soils Re-Engineering 131

8. Conclusion 132

References 135

LIST OF FIGURES

Fig. 1.1 Cross section of pavement under traffic cyclic loading..........3
Fig. 1.2 Cross section of pavement with crack propagation at subgrade failure under traffic cyclic loading...................4
Fig. 1.3 Plan of pavement with crack propagation at subgrade failure under traffic cyclic loading.................................4
Fig. 1. 4 Preparation sequence of Ceramics Waste Dust (CWD)..........6
Fig. 2.1. Solid Waste Sorting Process 8
Fig. 2.2. Crushed Solid Waste Production Procedure and Reuse for Ecofriendly, Ecoefficient and Sustainable Infrastructure...8
Fig. 2.3. Solid Waste Ash/Powder Utilization in Asphalt and Concrete Modification Process..9
Fig. 5.1 Grain Size Distribution of Studied Materials.....................22
Fig. 5.1.1 Effects of CWC on Consistency Limits of Treated Soils ..24
Fig. 5.1.2 Influences of Crushed Waste Glasses on Consistency Behaviour of treated soil: (a) Liquid Limits, (b) Plastic Limit, and (c) Plasticity Index 26
Fig. 5.1.3 Consistency Limits of QDbGPC Treated Soils................27
Fig. 5.1.4 Effect of CWP on the Consistency Limits of Treated Soils 29
Fig. 5.2.1 Effects of CWC on Compaction Characteristics of Treated Soils 32
Fig. 5.2.2 Influences of Crushed Waste Glasses on Compaction behaviour of treated soil: (a) Maximum Dry Density, (b) Optimum Moisture Content, (c) Specific Gravity 34
Fig. 5.2.3 Compaction Behaviour of QDbGPC Treated Soils 35

Fig. 5.2.4 Effects of CWP on Compaction Characteristics of Treated Soils 36
Fig. 5.2.5. Compaction of Test Soil A at 5% DOPC and Varying Proportions of Quarry Dust based Geopolymer under Varying Blows..................39
Fig. 5.2.5. Compaction of Test Soil A at 2.5% DOPC and Varying Proportions of Quarry Dust based Geopolymer under Varying Blows..................40
Fig. 5.2.6. Compaction of Test Soil A at 1% DOPC and Varying Proportions of Quarry Dust based Geopolymer under Varying Blows..................41
Fig. 5.2.7. Compaction of Test Soil A at 0% DOPC and Varying Proportions of Quarry Dust based Geopolymer under Varying Blow..................42
Fig. 5.2.8. Compaction of Test Soil B at 5% DOPC and Varying Proportions of Quarry Dust based Geopolymer under Varying Blows..................45
Fig. 5.2.9. Compaction of Test Soil B at 2.5% DOPC and Varying Proportions of Quarry Dust based Geopolymer under Varying Blows.................. 46
Fig. 5.2.10. Compaction of Test Soil B at 1% DOPC and Varying Proportions of Quarry Dust based Geopolymer under Varying Blows..................47
Figure 10. Compaction of Test Soil B at 0% DOPC and Varying Proportions of Quarry Dust based Geopolymer under Varying Blows..................48
Fig. 5.3.1 Effect of Quarry Dust Propotion on Swelling Potential Behaviour of Treated Test Soils: Soil A, Soil B, and Soil C 51
Fig. 5.3.2 Effect of Quarry Dust Propotion on Shrinkage Behaviour of Treated Test Soils: Soil A, Soil B, and Soil C..................52
Fig. 5.3.3 Effect of Quarry Dust Propotion on Swelling Potential Behaviour of Treated Test Soils: A, B, and C 53

Fig. 5.3.4 Effect of Quarry Dust Propotion on Shrinkage
   Behaviour of Treated Test Soils: A, B, and C................54
Fig. 5.4.1 Schematic Arrangement of the Sorptivity Test................57
Fig. 5.4.2 Effects of Quarry Dust on Sorptivity (a), and (b)
   Cumulative Infiltration of treated soil, A......................60
Fig. 5.4.3 Effects of Quarry Dust on Sorptivity (a), and (b)
   Cumulative Infiltration of treated soil, B......................61
Fig. 5.4.4 Effects of Quarry Dust on Sorptivity (a), and (b)
   Cumulative Infiltration of treated soil, C 62
Fig. 5.5.1 Effect of Quarry Dust Proportion on Erodibility
   Potential of Treated Soil... 64
Fig. 5.6.1 Effects of CWC on California Bearing Ratio
   Behavior of Treated Soils..66
Fig. 5.6.2 Effects of CWP on California Bearing Ratio of
   Treated Soils 67
Fig. 5.6.3 Effect of Crushed Waste Glasses Proportion on the
   CBR behaviour of DOPC+QDbGPC (%) treated soil..69
Fig. 5.6.4 California Bearing Ratio of OPC+QDbGPC (%)
   treated soil with CWG...70
Fig. 5.7.1 Unconfined Compressive Strength Behaviour of
   Treated Test Soils at 28 days curing: Soil A, Soil B,
   and Soil C..72
Fig. 5.7.2 Effect of Quarry Dust Propotion on Compressive
   Strength Loss on Immersed and Durability Index of
   Treated Test Soils: Soil A, Soil B, and Soil C................74
Fig. 5.7.3 Unconfined Compressive Strength Behaviour of
   Treated Test Soils at 28 days curing: A, B, and C.........75
Fig. 5.7.4 Effect of Quarry Dust Propotion on Compressive
   Strength Loss of Specimens Immersed and
   Durability Index of Treated Test Soils: 77
Fig. 5.8.1 Effect of CWC on Deformation (a) and R-Value
   behavior of Treated Soil A (b) 80
Fig. 5.9.1 Effect of CWC on Deformation and R-Value behavior
   of Treated Soil B 80

Fig. 5.10.1 Effect of CWC on Deformation and R-Value behaviour of Treated Soil C...81
Fig. 5.11.1 Effect of CWC on Deformation (a) and (b) R - Value behavior of Treated Soil D...................................81
Fig. 5.12.1 Effect of CWP on lateral Deformation (a), and (b) R-value behavior of Treated Soil A 83
Fig. 5.12.2 Effect of CWP on lateral Deformation (a), and (b) R-value behavior of Treated Soil B 84
Fig. 5.12.3 Effect of CWP on lateral Deformation (a), and (b) R-value behavior of Treated Soil C............................85
Fig. 5.12.4 Effect of CWP on lateral Deformation (a), and (b) R-value behavior of Treated Soil D............................86
Fig. 5.9a Effects of CWC on Deviatoric Stress of the Treated Cemented Soils 88
Fig. 5.9b Effects of CWC on Resilient Modulus, MR, of the Treated Cemented Soils 89
Fig. 6.4.3a Triangular simplex components 96
Fig. 6.5.1 Extreme vertices for; (a) 2-component simplex, 3-component simplex, 4- component simplex and 5-component simplex 105
Fig. 6.5.2. Factor space simplex of a 5- component mixture experiment for concrete production............................ 112
Fig. 6.5.4. Factor space simplex of a 4- component mixture experiment for asphalt production............................. 115
Fig. 6.5.6. Factor space simplex and contour space of a 3-component mixture experiment for soil stabilization .. 119
Fig. 6.5.8. Experimental simplex and factor space of the components in a 2- component mixture space............122
Fig. 6.5.9. Array factor space of the 5- component simplex of concrete production 124
Fig. 6.5.10. Trace and deviation factor space of the 5-component mixture for concrete production 125
Fig. 6.5.11. Array factor space of the 4- component simplex of asphalt production 126

Fig. 6.5.12. Trace and deviation factor space of the 4-component mixture for asphalt production 127

Fig. 6.5.13 Array factor space of the 3- component simplex of soil stabilization 128

Fig. 6.5.14. Trace and deviation factor space of the 3-component mixture for soil stabilization 129

Fig. 6.5.15. Trace and deviation factor space of the 2-component mixture for homogenous mixtures 130

LIST OF TABLES

Table 5.1 Basic properties of test soils 21
Table 5.6.1 California bearing ratio of OPC+QDbGPC (%) treated soil with CWG 70
Table 6.5.5.1. Design Matrix Evaluation for Mixture Quadratic Model 5 Factors: A, B, C, D, E 109
Table 6.5.5.2. Power at 5 % alpha level on 5- component for concrete production... 109
Table 6.5.5.3. Measures derived from the information matrix on 5- component for concrete production.................. 111
Table 6.5.5.4. Design Matrix Evaluation for Mixture Quadratic Model 4 Factors: A, B, C, D with U_Pseudo Mixture Component Coding; 112
Table 6.5.5.5. Power at 5 % alpha level on 4- component for asphalt production... 113
Table 6.5.5.6. Measures derived from the information matrix on 4- component for asphalt production.................... 114
Table 6.5.5.7. Design Matrix Evaluation for Mixture Quadratic Model 3 Factors: A, B, C with L_Pseudo Mixture Component Coding 116
Table 8. Power at 5 % alpha level on 3- component for soil treatment... 116
Table 6.5.5.9. Measures derived from the information matrix on 3- component for soil treatment 117
Table 6.5.5.10. Design Matrix Evaluation for Mixture Quadratic Model 2 Factors: A, B with L_Pseudo Mixture Component Coding;.............. 119

Table 6.5.5.11. Power at 5 % alpha level on 2- component for homogeneous mixtures 120
Table 6.5.5.12. Measures derived from the information matrix on 2- component 121
Table 6.5.7.13. 5- Component experimental mix proportions 123
Table 6.5.7.14. 4- Component experimental mix proportion 125
Table 6.5.7.15. 3- Component experimental mix proportions for soil stabilization 127
Table 6.5.7.16. 2- Component experimental mix proportions for concrete modification 129

PREFACE

As my undergraduate and postgraduate students, research associates and colleagues, practitioners, designers and constructors await the publication of this book later this year, it would have been four years after the conception of this book project aimed at bringing to our readers especially researchers and experts the ideas that have evolved through years of research, practice and teaching.

In a world faced with environmental issues most dreaded of which is global warming resulting from the depleting effect of oxides of carbon emission into the atmosphere, we have worked with assiduity to evolve alternative ways, sustainable enough to replace the utilization of ordinary Portland cement as a binder in foundation construction. The Nigerian General Specification recognizes these efforts. This book has included for the first time the blending of solid waste derivatives (ash and dust) with soft soils to enhance its engineering use and also the coupled forms of gematerials as geopolymer cements, which had served as supplementary to the conventional cement that has been in use for centuries.

Going further is the application of mathematical methods in the design of mixture experiment to optimize the properties of treated or re-engineered soils and further the monitoring of the behavior of the stabilized soils overtimes.

The fundamental features of this edition of this book is centered on achieving sustainable re-engineering of undesirable soils with ash and powder materials derived from solid waste which have shown to exhibit high pozzolanic behavior possessing high amounts of aluminosilicates.

It is very important to also note that direct combustion of solid waste to achieve ash also releases oxides of carbon but in this book, a model has been included to ensure efficient entrapping of the oxides of carbon in a controlled incinerator. This model is called the solid waste incinerator caustic soda oxides of carbon entrapment model (SWI-NaOH-OCEM). This model has proven to be a novel model that our readers can't wait to be abreast of.

This book has made provisions for graphical illustration of results achieved in the cause of teaching and research which places this work an edge over previous works in this subject. It combines the results of teaching and experimentations to help our readers' full understanding of the texts.

K. C. Onyelowe
April, 2019

NOTATION INDEX

CWD	crushed waste dust
CO	carbon monoxide
CO2	carbon dioxide
GPC	geopolymer cement
NaOH	Sodium Hydroxide (Caustic Soda)
$NaHCO_3$	Sodium bicarbonate (baking Soda)
Na_2CO_3	Sodium Carbonate (Soda Ash)
$H_2\uparrow$	hydrogen gas
SWI-NaOH-OCEM	Solid Waste Incinerator NaOH Oxides of Carbon Entrapment Model
PSA	particle size analysis
UV-VSA	Ultra Violet-Visual Spectrophotometer analysis
SEM	Scanning electron microscope
XRD	xray diffractometer
XRF	xray fluorescent
C-S-H	calcium silicate hydrate
C-A-H	calcium aluminate hydrate
OPC	ordinary Portland cement
Na_2SiO_3	sodium silicate
OH-	hydroxyl cation
H+	hydrogen anion
Ca++	calcium anion
Al+++	aluminum anion
Na++	sodium anion

K+	potassium anion
Mg++	magnesium anion
DDL	diffused double layer
Φ	repulsion potential
P	density of medium
Λ	dielectric constant
AASHTO	American Association State Highway and Transportation Officials
PI	Plasticity Index
LL	Liquid Limit
PL	Plastic Limit
SL	Shrinkage Limit
ASTM	American Standard for Testing and Mateials
AS	Australian Standard
IS	Indian Standard
NGS	Nigerian General Specification
GP	Poorly Graded
FSI	Free Swell Index
CH	High Clay
USCS	Universal Soils Classification System
MDD	Maximum Dry Density
OMC	Optimum Moisture Content
CBR	California Bearing Ratio
NMC	Natural Moisture Content
R-V	Resistance Value
M_R	Resilient Modulus
DOPC	Dangote Ordinary Portland Cement
CWC	Crushed Waste Ceramics
QDbGPC	Quarry Dust based Geopolymer Cement
QD	Quarry Dust
CWG	Crushed Waste Glasses
CWP	Crushed Waste Plastics

ψ_n	Potential
θ_n	Water content in sorptivity
D	soil diffusivity
S	Sorptivity
E_A, E_B, and E_C	erodibility potential
HBM	Hydraulically Bound Material
ANOVA	Analysis of Variance
V	moisture condition value
E	Subgrade Stiffness
n	degree of polynomial in scheffe's method
b	constant coefficient in scheffe's factor space
Z	actual components of scheffe
X	pseudo components of scheffe
X_1	fraction of water ratio in scheffe optimization
X_2	fraction of quarry dust or any additive for re-engineering
X_3	fraction of soils
L_b	lower bounds
U_b	upper bounds
XVERT	Extreme Vertices
Df	degrees of freedom
MATLAB	Matrix Laboratory

CHAPTER ONE

BACKGROUND

Soft soils encountered in construction works pose a big threat to the overtime performance and operation of facilities whose foundation or underlain structure is made of these materials. Pavements and pavement foundations are subjected to traffic loads like the form presented in Fig. 1.1, which could be axial or lateral pressure. The structures suffer lots of unacceptable behavior from construction cracks, volume changes to total collapse like the form presented in Figs. 1. 2 and 1. 3 if the foundation is such that the material is weak and lacks the sufficient strength to bear the loads. On the other hand, reengineering of soft clay soils is a procedure or method adopted by the experts in geotechnical engineering and soil mechanics to improve the properties of soils to meet minimum standard requirements. These minimum standard requirements are the conditions that justify the materials use as a foundation material or as a geomaterial. Soil reengineering involves mechanical, chemical and biochemical procedures. Through the application of these methods, soft soils achieve strength, density and durability. In recent years, the problem of choosing between which of these methods will give a more sustainable approach has been there. Earlier, Portland cement and other chemical additives and chlorides have been the only materials utilized as binders or pozzolans to improve on the mechanical properties of the soft soils or expansive soils. The utilization of recycled solid waste materials in crushed or amorphous forms as geomaterials has been proven technically reliable in the field of Geotechnical and Pavement Materials Engineering. Various waste materials have been in use in the treatment or stabilization procedure to improve the physical, mechanical, structural and geotechnical properties or characteristics of expansive or undesirable or stubborn test soils. Palm bunch ash, palm kernel shell ash, bagasse ash, periwinkle shell ash, paper ash, egg shell ash, rice husk ash, tire waste glass, etc. gotten by the direct combustion of the parent solid waste materials have been utilized in various ways and proportions to stabilize expansive soils for the purpose of foundation constructions and generally in composite forms for other geotechnical engineering uses. Results from these experimental protocols and exercises have shown that the test soils properties improved and satisfied basic construction requirements. Crushed oyster shells, crushed snail shell, crushed periwinkle shells, crushed waste glasses, crushed waste plastics, etc. have equally being used to stabilized test soils with great achievements. These crushed solid waste materials though biodegradable materials, exhibit pozzolanic properties that enhance their bonding strength with the treated soils hence improve their properties. In the case of ash materials, they exhibit amorphous and pozzolanic or cementitious properties serving as supplementary cementing materials. This makes them fit into the reactions that take place in the adsorbed complex or the double diffused layer interface where hydration, ions dissociation, calcinations, polymerization or geopolymerization, and cation exchange reactions take place. And by these processes, the formation of floccs within the treated blend is improved and eventual densification and strength gain. This is an important factor responsible for the subgrade stiffness or overall strength development of the treated soils. Upon the derivation of these ash materials during direct combustion, to be utilized in construction or the use of ordinary cement materials, there has been a build-up of the CO_2 emission into the atmosphere. This consequently contribute to global warming and eventually continue to put the environment at risk. In the case of conventional binder construction, for each ton of ordinary Portland cement used in construction works, an equivalent amount of CO_2 is released into the atmosphere. Research has been ongoing to search for an alternative for Portland cement. It has been discovered that most ash materials are pozzolanic in nature with very high content of aluminosilicates. So also are most crushed materials into powder form. These ash and powder materials are all derived from solid waste materials. For example, palm bunch ash, paper ash, periwinkle shell ash and snail shell ash, egg shell ash, bagasse ash, wood ash, rice husk ash, etc. are derived from palm bunch, waste paper, periwinkle shell, snail shell, egg shell, sugarcane fibre, wood, rice husk, etc. respectively. Again, oyster shell powder, quarry dust, snail shell dust, sawdust, crushed ceramics, crushed glasses, crushed plastics, etc. are derived from oyster shell, quarrying, snail shell, wood sawing, waste ceramics, waste glasses, waste plastics, etc. respectively. Research has shown that these materials contain high content of aluminosilicates hence highly pozzolanic. This means that they can replace Portland cement as alternative cementing materials or supplementary cementing materials. They are all derivatives of solid wastes and sustainably serve the purpose of soft soil re-engineering. A step further has been the synthesis of geopolymer cements with the derivatives of these solid waste materials as the base materials. Pavement facilities fail every day and this is caused by use of soils that are inadequate in strength and durability more especially in hydraulically bound conditions.

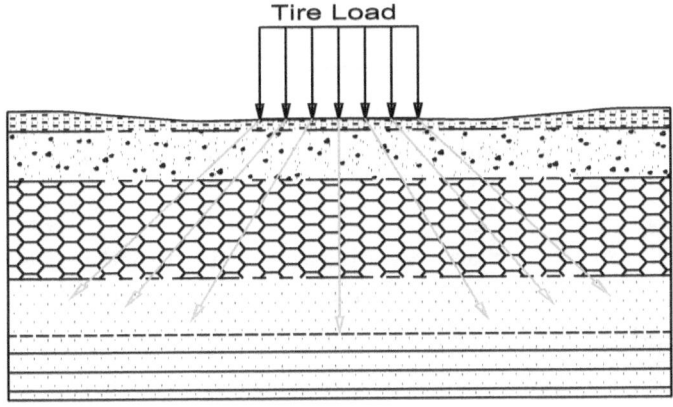

Fig. 1.1 Cross section of pavement under traffic cyclic loading

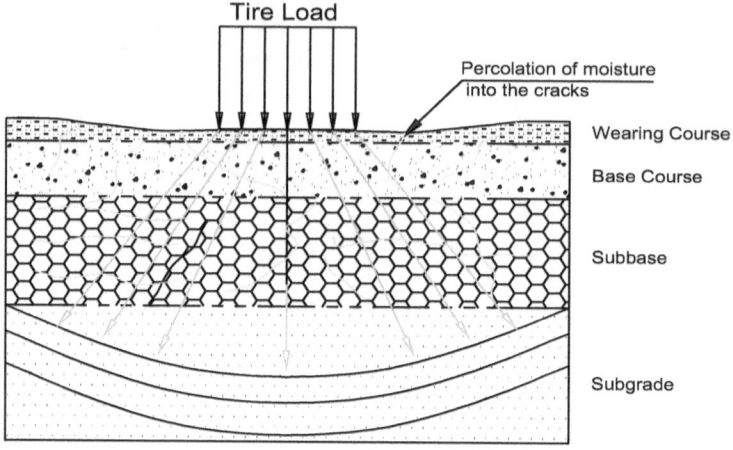

Fig. 1.2 Cross section of pavement with crack propagation at subgrade failure under traffic cyclic loading

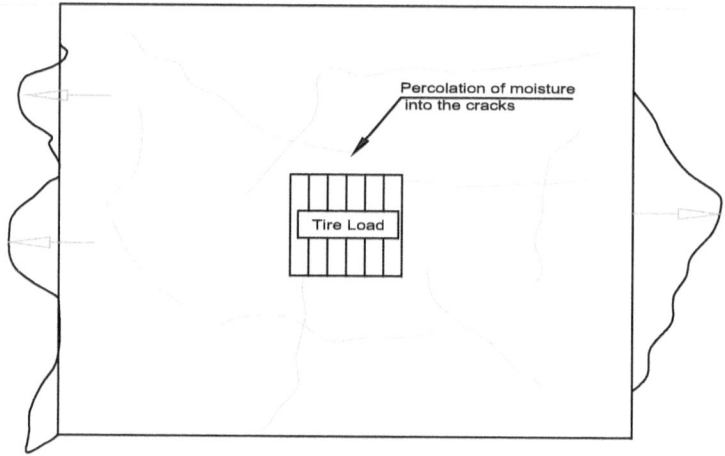

Fig. 1.3 Plan of pavement with crack propagation at subgrade failure under traffic cyclic loading

The utilization of crushed waste plastics and crushed waste ceramics has equally shown significant results in soft soil re-engineering. Plastics and ceramics are industrial products of polymerization and geopolymerization reactions, which possess elastoplastic properties. The discharge of waste from these products as household solid wastes and scrap losses from poor handling has been an environmental problem in the developing countries. These are recycled to serve various purposes, which included their utilization as geomaterials. Chemical oxides composition investigation has shown that these materials contain high level of aluminosilicates. These compounds are responsible for the stabilization reactions and geochemistry between test soils and admixtures. The disposal of ceramic wastes in landfills poses a potential hazard and threat to the environment and humans. Ceramic wastes are industrial wastes released into the environment through poor industrial and business handling of these materials. They are also released as household wastes and these are disposed indiscriminately without recourse to its potential dangers. They are derivatives of tiles, plates, cups, artifacts, and various other materials made from the high aluminosilicate composite material. Research has shown that ceramic wastes exhibit some structural and mechanical properties, which make them good fits as construction materials and more especially as geomaterials applied in soils re-engineering. Structural engineers have used significant amount of these wastes in the production of concrete and produced favorable results that met the minimum requirements for its utilization as a fifth component of concrete mixes. The utilization of these waste materials in the production of concrete has created a fulcrum for its disposal by recycling and reuse as a construction material. Further in this effort is the utilization of ceramic waste dust derived from the crushing of ceramic wastes (See Fig. 1.4) as a geomaterial applied in soft soil mechanical and structural properties improvement and re-engineering. Ceramic materials have been observed as possessing high density, abrasive strength, compressive strength, frictional potential, hydration potential and sulfate and heat resistance properties. These properties are beneficial in the construction of foundations in moisture bound environments where structures are almost constantly exposed to moisture throughout the life of the facility. The most important factor considered in hydraulically bound constructions like substructures is the volume changes; swell and shrink cycle soft clay soils experience when exposed to moisture. Volume changes exhibited by soft clay soils used as subgrade materials in pavement foundation cause most of the structural failures observed on pavement facilities, which include highways pavements, airfield, parking lots, platforms, etc. When subgrade layer material swells or shrinks beyond the allowable limits, frosts, heaves, cracks, fractures, etc. start to form on the paved facilities. This further gives way for further migration of moisture down the foundation structure and the eventual failure and collapse of the structure resulting from moisture effects.

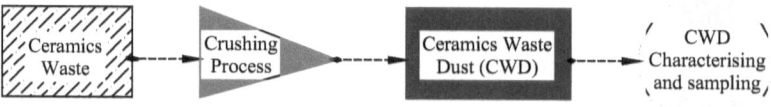

Fig. 1. 4 Preparation sequence of Ceramics Waste Dust (CWD)

This book is tentatively made up of eight chapters. Chapter one presents are background of the content of this book. Chapter two tries to present an idea of what to look out for in the entire book on sustainability of soils re-engineering in a 21st century geotechnical engineering practice. Chapter three brings to bear on the sustainable materials and practices that have given rise to the achievements made in the utilization of derivatives of solid waste materials re-engineering soils for the purposes of foundation engineering works. Chapter four deals on the soils and materials blending and the promises of this technique. Chapter five presents the achievements made on the improvement of the engineering properties of soils with the blending and stabilization protocols on soils. Chapter six presents the mathematical methods that have been employed in the design of experimental mixes and blending of soil and admixture combination and also employed in monitoring the behavior of treated and stabilized foundation materials. Chapter seven presents in general environmental soils re-engineering and finally Chapter eight presents an executive conclusion of the entire presentation.

CHAPTER TWO

SUSTAINABILITY IN SOILS RE-ENGINEERING

Sustainability in soils re-engineering in the 21st century geotechnical engineering practice is centered on oxides of carbon emission free construction practices and the utilization of the derivatives of solid waste. Solid waste handling and management around the world and more especially in the developing countries have contributed to the whooping amounts in tons of carbon dioxide (CO_2) and carbon monoxide (CO) emission and consequently on the threat of global warming. Lots of human activities are involved in this negative contribution to environmental hazard including; construction activities that utilize Portland cement, industrial activities that release oxides of carbon and other volatile gases, agricultural activities that release biomass and biopeels, mechanical activities for example engine-fuel combustion, which releases CO/CO_2, etc. The direct combustion or crushing, after they have been carefully sorted (see Fig. 2.1), of agricultural, household, municipal and certain biomass materials is not left out in this list.

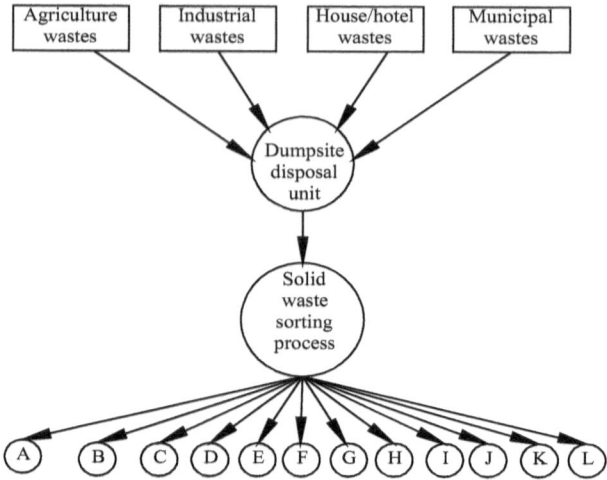

Fig. 2.1. Solid Waste Sorting Process

All this activities release CO and CO_2 into the atmosphere, which contributes over 80% of the worlds volatile and hazardous emissions. More important to deal with in this book is the recycling and reuse of solid waste materials by sorting, burning to ashes or crushing as the case may be and the utilization of the product of this process (ash or powder) in various ways in construction works (See Fig. 2.2).

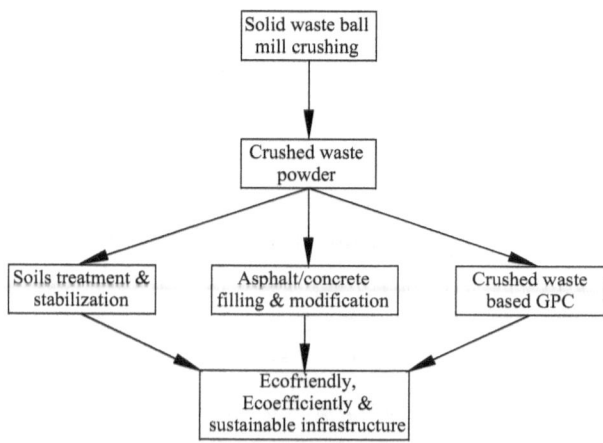

Fig. 2.2. Crushed Solid Waste Production Procedure and Reuse for Ecofriendly, Ecoefficient and Sustainable Infrastructure

A certain class of solid wastes known as bio-peels are also discharged into the landfills and they decay emitting volatile and dangerous gaseous compounds, which eventually impress on the environment and humanity in general. Very many others belong to the group of wastes that are also discharged on landfills and this poses a serious threat to the environment. In recent times, there have been strong calls for experts in engineering to work and exercise their expertise and practice towards ecofriendly, eco-efficient and more sustainable operations and designs to protect our planet from the dangers of global warming. Ordinary Portland cement as it is known, contribute equivalent amount of CO_2 emission into the atmosphere as it has found its use in almost every civil engineering construction exercise. It is also known that cement has its low ebb/performance in terms of resistance to cracking, durability issues (long-term exposure to moisture), brittleness, fire and heat resistance and sulfate resistance. On the other hand, solid waste materials derivatives like ash (amorphous) and powder/dust (crushed) have been discovered as possessing properties that not only replace cement in parts or whole, but improve on the resistance to sulfate attacks, heat, fire, suction, capillary action, sorption, cracking, etc. of infrastructures where they are adapted (See Fig. 2.3).

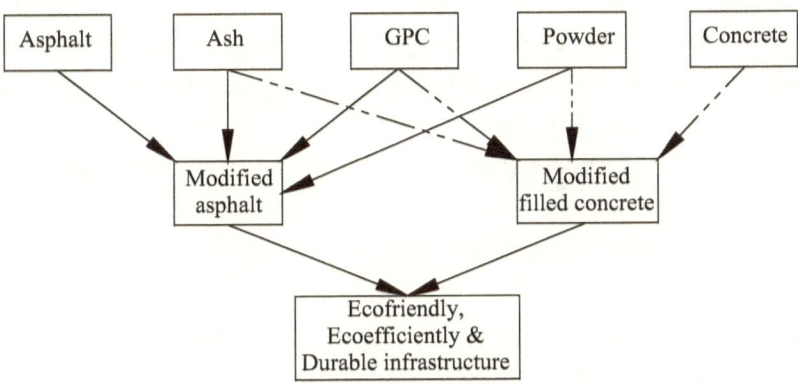

Fig. 2.3. Solid Waste Ash/Powder Utilization in Asphalt and Concrete Modification Process

Further on, these materials because of their high pozzolanic properties or high aluminosilicates contents have found use in the synthesis of geopolymer cements utilized in various proportions in soft soils reengineering, asphalt and concrete modification. It is also important to note that certain solid waste ashes have found their engineering use in the production of composite machine parts from alloys of Aluminum, Magnesium, Copper, etc. utilizing solid waste as reinforcements. These operations are all ecofriendly and contribute nothing to the menace of global warming and the infrastructures arising from this have been shown by experimental results as eco-efficient in life service, performance and durability. Also these environmental friendly binders (geopolymer cements) are derived from solid wastes under the influence of alkali activators. Inasmuch as it is certain that human activity will never cease on the planet, there would always be an equivalent release of solid wastes (industrial, agricultural, household, and municipal, etc.). Hence sustaining this technology wouldn't be a problem at all. Among the numerous derivatives of solid waste that have been utilized in various civil engineering works include palm oil fuel ash, fly ash, quarry dust, rice husk ash, coffee husk ash, paper ash, waste tire ash, palm fiber, palm kernel shell ash, snail shell ash, periwinkle shell ash/powder, oyster shell powder, biomass ash, bagasse ash, egg shell ash, sawdust, crushed waste ceramics, crushed waste plastics, crushed waste glasses, bio-peels, metallurgical slag (ground granulated blast furnace slag), iron ore tailings, palm nut fiber, glass fibers, full depth reclaimed asphalt, etc. It is important to note that this book is aimed at bringing to our readers the extent to which these practices can be sustained with low or zero carbon generation and emission. Our planet continues to be endangered with these emissions from agricultural, industrial, household and municipal processes and requires emergency reversal of those activities contributing to its depletion.

CHAPTER THREE

SUSTAINABLE MATERIALS PROCESSING

3.1 Processing Ash by Direct Combustion

The greatest amount of ash materials utilized in soil stabilization, concrete production, asphalt modification and the synthesis of geopolymer cements (GPC) are derived through direct combustion in well-designed incinerators. In most cases, this combustion exercise is uncontrolled. This by implication poses the greatest danger of CO and CO_2 emission to the environment. Like in the case of crushed solid wastes, the selected wastes are collected, sorted and burnt in an incinerator for a controlled combustion. In this case, a model of the incinerator has been designed to ensure that the CO and CO_2 released through the firing smoke is entrapped. This mechanism is to ensure that the whole essence of an ecofriendly operation in civil engineering works is not defeated. The utilization of ash or its forms in constructions and as geomaterials for the replacement of ordinary Portland cement is to reduce to zero those construction practices that release oxides of carbon into the atmosphere to ensure an environmental friendly activity. But the production of ash by combustion is against this aim. What has been done was to use the caustic soda-incinerator model to trap volatile gases released during the combustion process. The model is such that ensures that CO and CO_2 release during solid waste combustion is entrapped by caustic soda solution (prepared NaOH, 40% w/v). 100% of the CO and CO_2 released is trapped by the caustic soda solution because of the affinity caustic soda has with oxides of carbon. This entrapment produces sodium bicarbonate (baking soda, $NaHCO_3$), sodium carbonate (soda ash, Na_2CO_3) and hydrogen gas (H_2) presented in Equations 3.1 and 3.2;

$$NaOH + CO_2 \ggggggggggg NaHCO_3 \quad (3.1)$$
$$NaOH + CO \ggggggggggg Na_2CO_3 + H_2\uparrow \quad (3.2)$$

The caustic soda (sodium hydroxide) entrapped oxides of carbon that would been released to the environment with their dangers have been converted to household and industrial compounds. The use of baking soda ($NaHCO_3$) in households and industries cannot be overemphasized. So also is the use of soda ash (Na_2CO_3) in the synthesis of geopolymer cements; a geomaterial binder with great properties. Additionally, small amounts of the soda ash improves the flocculation properties of treated soft soils and treated wastewater by improving the binding of fine particles. Hydrogen gas is also a product of the Solid Waste Incinerator NaOH Oxides of Carbon Entrapment Model (SWI-NaOH-OCEM). This gas and its isotopes have many uses in the field of science and atomic physics. It is important to note that the products of this model; SWI-NaOH-OCEM are water soluble compounds and element and find beneficial uses to man and environment. However, ash which is the primary product of the SWI-NaOH-OCEM has been in several ways in construction works. Research results show that because of its amorphous nature and high aluminosilicates content, it acts as a binder with high pozzolanic property. It has several cases consistently improved the engineering properties of expansive soils, concrete, and asphalt.

3.2 Processing Powder by Direct Crushing

The utilization of various forms crushed solid waste in the reengineering of soil, concrete, wastewater, and pavements has been made possible by the ball mill crushing of solid waste to obtain crushed waste plastics, crushed waste ceramics, crushed waste glasses, crushed oyster shell powder, etc. The selected solid waste materials are collected from dumpsites, sorted, washed and sundried. They are thereafter crushed with the ball mill to different sizes and texture. Characterization of these crushed powder materials is also done by particle size analysis (PSA) and UV-Vis Spectrophotometer analysis (UV-VSA). In more advanced technologies, the Scanning Electron Microscopy (SEM), X-ray Fluorescent (XRF) or the X-ray Diffractometer (XRD) is used to determine through the plotted diffractograph, the micro properties and gradation behavior of the crushed materials. This also indicates the overall performance of the materials in blends with soil, concrete or asphalt. These crushed derivatives of solid waste have proven through experimentation to be good binders and fillers and modifiers in soil stabilization, concrete production and asphalt production respectively. These materials have been added in varying percentages by weight of treated homogenous mixture (soil, concrete or asphalt) to improve their engineering properties. This has been possible because these materials were discovered to possess pozzolanic properties that enhance cementation processes. Their ability to form compounds responsible for strengthening was as a results of high content of aluminosilicates, which makes it possible to calcium silicate hydrates (C-S-H) and calcium aluminate hydrates (C-A-H) with the treated mass or mixture.

3.3 Processing Coupled Materials by Blending

Environmental friendly Geopolymer cements are coupled binders synthesized by coupled-blending of solid waste ash or crushed waste ash under the activation influence of alkali activators. These are developed to attend to the urgent need for ecofriendly and eco-efficient binders and to totally or partially replace conventional cements (OPC) in all the civil engineering works to save our planet from the hazards of CO/CO_2 emission. The alkali activators are produced under the laboratory conditions with Sodium Hydroxide (NaOH) and Sodium Silicate (Na_2SiO_3) as activators with a NaOH molar concentration of 12M for an eco-friendly material. A proportion of 4.8% by weight activator is used to blend the selected ashes. Research results have shown that various forms of geopolymer cements have been synthesized from this operation which included quarry dust based geopolymer cement, crushed waste ceramics geopolymer cement, crushed waste glasses geopolymer cement, palm bunch ash based geopolymer cement, bagasse ash based geopolymer cement, etc. These forms of GPC have shown to possess similar properties in terms of performance and behavior in their utilization as alternative or supplementary binders. The utilization of the GPC has shown to produce infrastructures with certain special properties. Results have shown that GPCs possess high potentials to resist sulfate attacks, moisture attacks in hydraulically bound environments, heat and high temperature, volume changes in soft clay and expansive soils, cracking in concrete structures, lateral displacement in asphalt pavement, capillary action, sorption and suction in subgrade materials and generally to improve the durability of civil infrastructures. In the run, the eco-efficiency of the facilities from these ecofriendly materials and geomaterials.

CHAPTER FOUR

SOILS AND MATERIALS BLENDING

4.1 Soils, Materials and Proportionate Mixing

Geotechnical engineering works depend on soil for sustainable infrastructural and sub structural developments. Years of practice and teaching experience have shown that most engineering soils for use as foundation materials fall below desirable characteristics and requires improvement in terms of usage and technical need. Hence soils are collected and characterized for engineering classifications necessary for certain and various engineering use. Soils are utilized as components in concrete production and more especially in blends of geomaterials for subgrade of airfields pavement, highway pavements, parking lots pavements, and various others. So, in geotechnical engineering works, soil is the primary component that requires re-engineering to meet engineering construction requirements. It is also the determinant of the proportion of other admixtures during the treating protocol. Hence other materials of ash or powder derived through direct combustion or crushing are proportioned by weight of the treated or improved or stabilized soil as the case may be. For far too long, the environment has suffered the hazardous effects of CO and CO_2 emission through human and engineering activities, which deplete the ozone layer and its attendant contribution to global warming. Expert of climate change in different fora and congresses and summits have given to the world what CO/CO_2 emission is doing to our planet. Players in this menace destroying our planet have been urged to seek for ways and alternative processes to renew our planet through environmental friendly activities and practices. The utilization of Portland cement in civil engineering works seems to contribute a large amount of the oxides of carbon to the atmosphere because has shown that ordinary Portland cement (OPC) use releases equivalent of amount of CO_2 into the environment, that is to say that one ton of OPC used in any construction activity release an equivalent one ton of CO_2 into the atmosphere. This is shocking and dangerous and if these activities continue, the future of this planet seems gloomy and uncertain. OPC is used in civil engineering work for the reengineering of soils and concrete as structural and pavement foundation materials but this has failed both the ecofriendly dream environment by releasing CO_2 and eco-efficient product by failing durability tests because cemented structures are vulnerable to heat, sulfate, moisture, crack and shrinkage effects. On the other hand, ecofriendly materials that replace OPC as supplementary cementing materials are usually derived from direct combustion, another procedure that releases CO/CO_2 during the burning process. What this research has achieved among the ash derived from solid waste direct combustion, is trapping the CO/CO_2 released during solid waste combustion using the SWI-NaOH-OCEM. From this model, we have the ash materials adapted to many different forms of geomaterials and construction materials, CO/CO_2 entrapment and release of environmental friendly products ($NaHCO_3$, Na_2CO_3 and H_2) for household and industrial uses. Away from the ecofriendly disposition in handling and utilizing ash as a construction material, the treatment of soil, concrete, asphalt, etc. with ash and its products (geopolymer cements) produces structural infrastructures with high resistance to heat, temperatures, salts attacks, crack effects, punching deformations, lateral displacements, rise and fall of water tables, which give rise to volume changes (swelling & shrinkage), moisture and capillary actions and brittleness. These structures also possess higher durability potentials as shown by previous research results. Human, agricultural and industrial will always be parts of this environment hence the release of solid waste is a corresponding certainty. However the sustainability of the SWI-NaOH-OCEM is never in doubt and consequently the sustainability of the derivatives of solid wastes.

4.2 Diffused Double Layer Reactions and the Adsorbed Complex

Adsorbed moisture is the moisture located within the influence zone of the contact between water and clay particles whose properties are very viscous and different from those of free or normal water at the same temperature. The behaviour of soil matrix depends upon the behaviour of the discrete elements composing the matrix and the structural pattern of the particles organization. In all these states, moisture plays a very important role as it is applied in soil stabilization at different moisture contents and dry densities. The reaction of the soil matrix is profoundly influenced by the inter-particle–water relationships, the ability of the soil elements to adsorb exchangeable cations and the amount of moisture present and free to react, which also depends on the orientation of the net ions within the clay matrix and the dipolar water. The clay particles carry negative charge on their surface and this is the result of isomorphous substitution and of a discontinuity of the structure at its edges. As it is mentioned above, cation exchange is a process by which stronger cations in the metallic elemental series displace weaker ions. Electrolytes dissociate when dissolved in moisture into positively charged cations and negatively charged anions like water which dissociates into hydrogen cation $H+$ and hydroxyl anion $OH-$. These positively charged ions migrate to the surface dominated by the negatively charged clay particles and form the adsorbed layer. From the metallic order of the ions, these $H+$ ions can be replaced by the other cations such as $Ca++$, $Al+++$, $Na++$, $K+$, or $Mg++$. When these ions migrate into the adsorbed layers, they constitute the "adsorption complex". This process of replacement of cations of one kind by those of stronger kind within the adsorption complex is known as "Base Exchange", which is meant the capacity of colloidal particles to change the cations adsorbed on their surface. This takes place by a constant percolation of water containing dissolved sodium salts. And the quantity of the exchangeable cations in a soil matrix is called "exchange capacity". The above illustrations are the mechanics of the interaction that take place in any soil stabilization operation. In Geotechnical engineering works, weak soils call for a need to carry out stabilization to chemically or mechanically change the properties of such soils to make them useable and serviceable to the Geotechnical engineer. And soils used in Geotechnical engineering services are made primarily of clay. Clay minerals as are contained in lateritic soils are complex aluminium silicates compounds of one of two basic units which are tetrahedral units of four oxygen atoms surrounding a silicon atom forming silica sheets and the octahedral units of six hydroxyl compounds surrounding an aluminium atom forming gibbsite sheets or surrounding a magnesium atom forming brucite sheets. These are the major building block of clay minerals. When soils are subjected to stabilization by a combined effect of chemical, mechanical and admixture procedures and processes, ions are released within the interface between the clay particles and the additives. The dominant negative ions from clay are balanced by exchangeable cations like Ca^{++}, Mg^{++}, Na^+ and K^+ from the additives surrounding the particles being held by electrostatic attraction and the van der Waal's force. The release of the ions on the other hand depends also on the oxide composition of the additives. At the point that moisture is added to the soil matrix being stabilized, these cations and a small number of anions float around the clay particles within the diffused double layer (DDL). Research has shown that cation concentration decreases with the distance from the surface of the particles while the anion concentration increases but slowly. For flocculation, densification and strength gain to be complete in a stabilization operation, there must be a comprehensive cation exchange which is to say that the distance between the particles has to be reduced to improve the cation concentration. If we assume that the ions in the double layer can be treated as point charges and that the surface of the stabilized clay elements is large compared to the thickness of the double layer, Poisson's equation states,

$$\frac{d_2 \emptyset}{dx^2} = -4\pi\rho \tag{4.1}$$

Where Φ = repulsion potential, ρ = density of medium, λ = dielectric constant and the x = distance between charged particles and clay surface. The repulsion potential is the tendency for the distance between the particles which we consider as charged points to increase thereby reducing the potentials for the stabilized matrix to form flocs and eventually gain strength and density.

CHAPTER FIVE

SUSTAINABLE SOILS PROPERTIES IMPROVEMENT

5.1 Gradation and Characterization of Selected Soils

In this section a practical case study of four soil samples were prepared and preliminary tests were also conducted to ascertain the basic geotechnical and mechanical properties of the soils. This was to enable proper classification and characterization for use as construction materials. The basic experimentations were carried out in accordance to American Association of State Highway and Transportation Officials, American Standard for Testing and Materials, British Standard International, Indian Standards, Nigerian General Specifications and Australian Standards. This is done to establishment international usage and replication of the methods and results as presented in this book. The basic properties of the test soils are presented in Table 5.1 and Fig. 5.1. The test soil were observed to possess 2.85%, 10%, 4.6% and 7.6% passing sieve No. 200, and classified as A-2-7, A-2-6, A-7 and A-7-5 respectively according to AASHTO classification method. Test soils A, C and D were also classified as poorly graded with high clay content while test soil B was classified as poorly graded according to unified soil classification system. The results of the consistency protocol show that the test soils are highly plastic soils (PI > 17%) with high free swell index. The basic results of the resistance (r) value indicates that the test soils are sandy clay group (10-20) while that of the resilient modulus shows that the soils fall under clayey subgrade (0.345E+05 to 1.034E+05 kN/m²). With soils of such poor characteristics to meet the basic requirements for use as foundation materials, it is important that experimentation for improvement and stabilization were carried. This would enable the soils to be utilized as, not just subgrade formation but as base materials, backfills, waste containment liners, and other geomaterials purposes and even as high density concrete materials. We would take a long walk to the improvements that have been made over the years trying to work on and re-engineer the soft clay, expansive and undesirable soils to fit into the engineering works standard and requirement. It will benefit undergraduate students in civil and mechanical engineering that may someday be working on composites reinforced with ash materials derived from solid waste. More to benefit from the resources contained in this book are also research and graduate students, practitioners and contractors. Let's take a look!

Table 5.1 Basic properties of test soils

Property description of test soils and units	Values			
	Test soil (A)	Test soil (B)	Test soil (C)	Test soil (D)
% Passing Sieve No 200	2.85	10	4.6	7.6
NMC (%)	12.1	13.49	14	16
LL (%)	40	46	64	65
PL (%)	18	21	36	33
PI (%)	22	25	28	32
SL (%)	8	8	7	10
FSI (%)	250	234	275	296
G_s	2.6	2.43	2.12	2.08
AASHTO Classification	A-2-7	A-2-6	A-7	A-7-5
USCS	GP, CH	GP	GP, CH	GP, CH
MDD (g/cm³)	1.76	1.85	1.80	1.56

OMC (%)	13.1	16.2	13.13	15.4
CBR (%)	12	13	8	7
R-Value	11.74	11.70	11.70	11.50
M_R (kN/m²)	0.42E+05	0.42E+05	0.42E+05	0.72E+05
Color	Reddish Brown	Reddish Gray	Reddish Ash	Ash

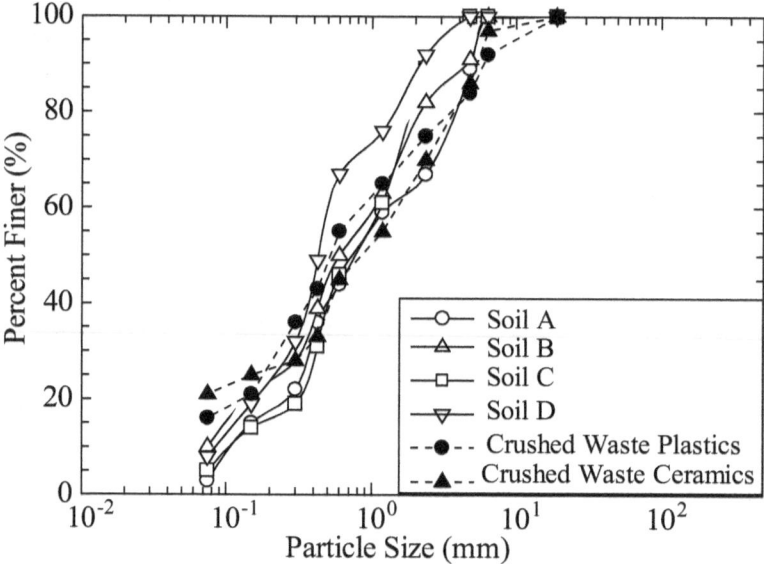

Fig. 5.1 Grain Size Distribution of Studied Materials

5.1 Consistency

We shall be using different case studies to send home our ideas and laboratory discoveries in terms of re-engineering processes carried out on soils. Because various situations arise in the field of geotechnical engineering and geo-environmental practice where soils form a central material.

The first case is where consistency behavior of 2.5% addition of dangote ordinary Portland cement (DOPC) and varying proportions of crushed waste ceramics (CWC) treated test soils A, B, C and D was presented in Fig. 5.1.1. Preliminary results had shown that the test soils were classified as highly plastic soils with high clay content responsible for the change in volume by swelling and shrinkage under the influence of moisture. This property in effect shows that the soils are expansive exhibiting poor consistency properties. The addition of the crushed waste ceramics in the treatment exercise showed that the test soils decreased consistently in their plasticity value. The test soils improved from highly plastic soils to very low plastic soils with highly stiff consistency with further and incremental addition of the CWC. The behavior had resulted from various reasons; the release of cations and the release of carbons by the highly aluminosilicates of CWC at the adsorbed complex zone of the mixed treated soils. These released materials in different states ionization were responsible for the hydration reaction at the initial phase of the treatment protocol. This led to the dissociation of ions under the laboratory moisture content. Subsequently, carbonation and calcination reaction took place at the reaction interface of the reactive minerals from soils and the additives. At the stage, the reactive minerals form floccs with the treated soils giving rise to densified and strengthened blend of treated material. This observed and recorded improvement was due to the hydration of the blend of the highly pozzolanic admixture and the treated soils giving rise to a reduced plasticity index. This equally brought about the formation of the stiff consistency of the treated matrix. The liquid and plastic limits equally reduced at the addition of the CWC to the treated soils. This showed that the moisture content was dependent on the physicochemical properties of the added admixture. This behavior agrees with research results carried out by the first author of this book. This states that as water is used as pore fluid, the influence of the mechanical factors would remain same with an overall reduction in liquid limits of the treated soils on addition of an additives. The addition or treatment of the test soils with the crushed waste ceramics not only achieved a stiff consistency material, but achieved non-frost-susceptible materials with PI less than 15. This behavior is responsible for the pavement failures resulting from the formation of frosts. With the addition of CWC and reduction of plasticity index below 15, the exercise achieved are more durable treated matrix as subgrade foundation material.

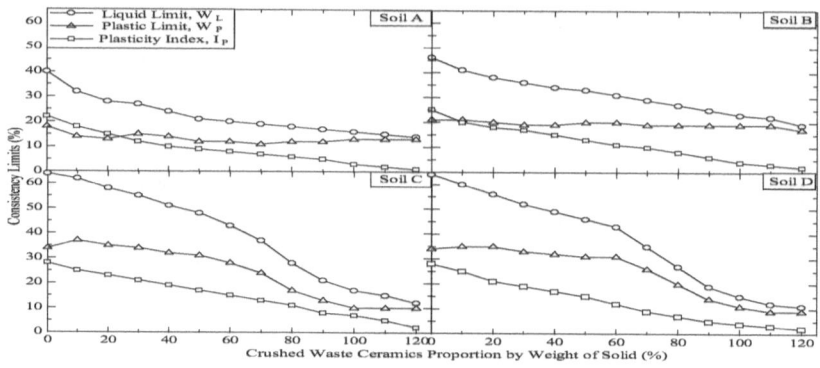

Fig. 5.1.1 Effects of CWC on Consistency Limits of Treated Soils

The second case is the consistency behaviour of a coupled mixture of quarry dust based geopolymer cement (QDbGPC) to Dangote Ordinary Portland Cement treated soil A with crushed waste glasses (CWG). Fig.5.1.2 present the consistency behaviour of the quarry dust based geopolymer cement linearly and inversely replace ordinary Portland cement treated soil under laboratory conditions under crushed waste glasses added to the treatment procedure. The addition of CWG into the cemented soil reduced the liquid limits, plastic limits and the plasticity index consistently. This behaviour however shows that further addition of CWG beyond the maximum 40% utilized in this exercise could have improved the consistency limits further. But very important to note was the improvements recorded with the linearly inversely introduction of the cements into the test soil. That is, while the bio-based geopolymer cement was increased in the treatment blend, the Portland cement was reduced and the effect of this treatment pattern was observed. Results have shown that increased quarry dust based geopolymer cement reduced the consistency limits from very high plastic condition to even very less plastic consistency. The hydration of the stabilized mixture and its increased calcination and pozzolanic activity have contributed to the behaviour of the soil. And also due to molecular rearrangement in the formation of transitional compounds. This improvement is due to the hydration of the highly pozzolanic additives from the quarry dust based geopolymer cement (QDbGPC) with the treated soil matrix, which reduced the PI consistently thereby producing a stiff mixture of stabilized soil. Also, the release of more cations from the biomass based geomaterials and quarry dust during the cation exchange reaction has contributed to the behaviour of the stabilized mixture. This behaviour agrees with Meegoda & Ratanweera, which showed that if water is used as pore fluid, the influence of the mechanical factors would remain same with a general decrease in LL and PI on addition of an admixture and binder. The prone to cracks and brittle behaviour of Portland treated soils has contributed to the improved consistency limits at reduced rates of the DOPC.

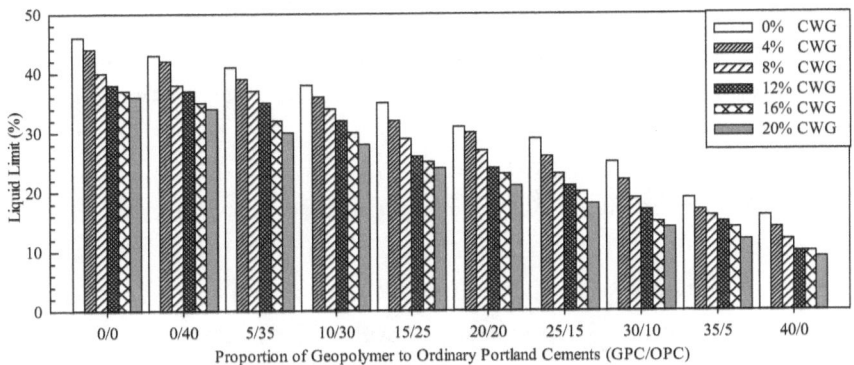

(a)

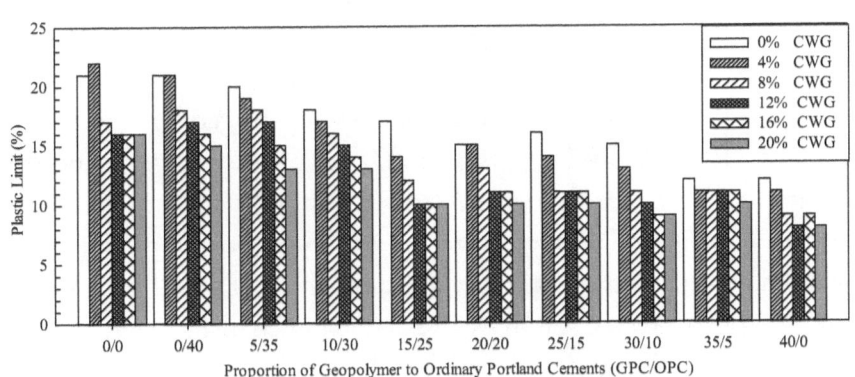

(b)

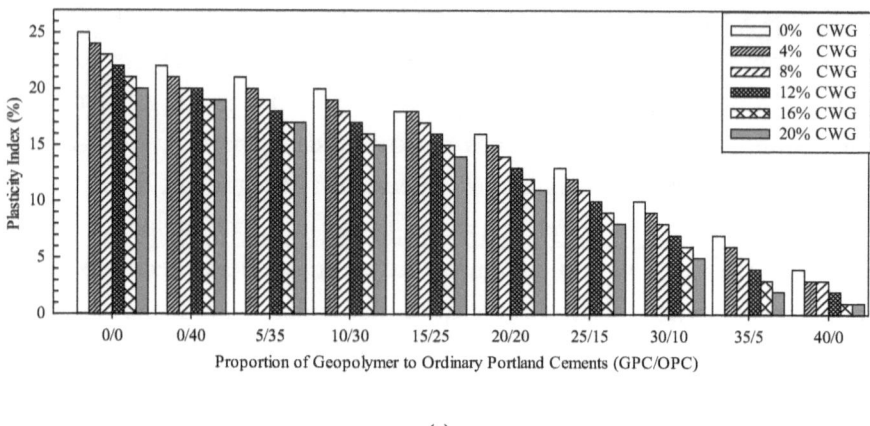

(c)

Fig. 5.1.2 Influences of Crushed Waste Glasses on Consistency Behaviour of treated soil: (a) Liquid Limits, (b) Plastic Limit, and (c) Plasticity Index

The third case is the consistency behaviour of quarry dust based geopolymer cement (QDbGPC) treated uncemented lateritic soils are presented in Fig. 5.1.3. In the case, ordinary Portland cement was eliminated entirely and the coupled material was used for the re-engineering process on three selected soils. The plasticity indexes (PI) of the natural untreated lateritic soils A, B, and C were recorded as 22, 25 and 28% respectively. These were considered highly plastic soils and undesirable as construction materials. Upon the treatment of these soils with QDbGPC, the PI was observed to reduce consistently at the rate of 18%, 8% and 4% respectively at 10% by weight utilization of the geopolymer cement. On further treatment with the geopolymer cement at 20% to 100% by weight utilization, the PI consistently reduced at the rate of 9% for treated soils A and B and 13% for treated soil C. At 130%, 140% and 150% utilization of the geopolymer cement, the consistency behaviour of soil A showed a steady PI of 4%. This observed behaviour of the treated soils have been contributed by increased calcium content from the geopolymer cement blend to hydration reaction and calcination. This was also due to the rearrangement of molecules during the formation of transitional compounds. The hydration and calcination of the treated soils under high proportional aluminosilicate and pozzolanic geopolymer cement had caused this improvement. This however, produced stabilized treated soils of stiff consistency. Moreover, the cations release deposited at the adsorbed complex interface of the treated soils from the synthesized geopolymer cement constituents during the cation exchange reaction had also caused this reaction. This proves that the mechanical factors of the treated soils would remain steady with the usual reduction in the liquid limits on addition of the treatment material if water is used as pore fluid. With the foregoing, under the addition of varying proportions of quarry dust based geopolymer cement, it can be deduced that the liquid limits and plasticity indexes depend on the mechanical conditions rather than the viscosity of pore fluid and density of the treated matrix. But to a higher degree, this treatment protocol results is a function of carbonation, cation exchange, polymerization and polycondensation, which are physicochemical properties. Having achieved treated materials matrix of plasticity index below 15%, the utilization of quarry dust based geopolymer cement has improved the treated lateritic soils such that they can best be used at non-frost-susceptible subgrade and subbase materials. That is to say that under hydraulically bound or moisture bound conditions, the materials can adjust with great flexibility to withstand the effect of that exposure.

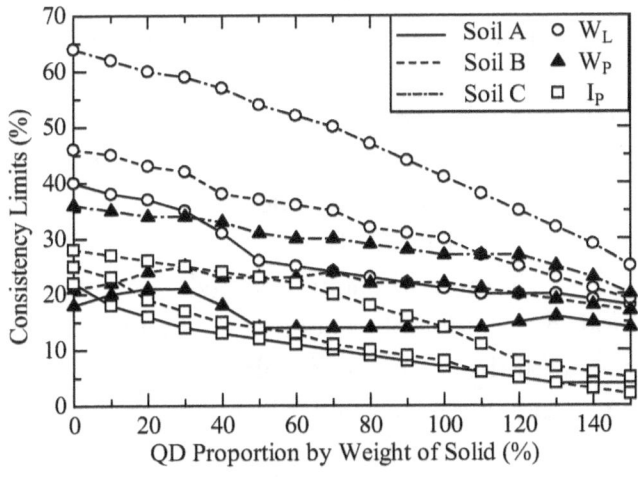

Fig. 5.1.3 Consistency Limits of QDbGPC Treated Soils

The fourth case is the consistency behavior of 2.5% DOPC and varying proportions of crushed waste plastics (CWP) treated test soils was presented in Fig. 5.1.4. Preliminary results had shown that the test soils were classified as highly plastic soils with high clay content responsible for the volume changes in moisture. This property in effect shows that the soils are expansive exhibiting poor consistency properties. The addition of the crushed waste plastics (CWP) in the treatment protocol showed that the test soils decreased consistently with the plasticity value. The test soils improved from highly plastic soils to very low plastic soils with highly stiff consistency with further and incremental addition of the CWP. The behavior had resulted due to; the release of cations exchange by the highly aluminosilicates of CWP at the double diffused layer (DDL) zone of the blended treated soils. These released materials in different states of ionization were responsible for the hydration reaction at the initial phase of the treatment protocol. This led to the dissociation of ions under the laboratory moisture content. Subsequently, carbonation and calcination reaction took place at the reaction interface of the reactive minerals from soils and the additives. At this stage, the reactive minerals form floccs with the treated soils giving rise to densified and strengthened blend of treated material. This observed and recorded improvement was due to the hydration of the blend of the highly pozzolanic admixture and the treated soils giving rise to a reduced plasticity index. This equally brought about the formation of the stiff consistency of the treated matrix. The liquid and plastic limits equally reduced at the addition of the CWP to the treated soils. This showed that the moisture content was dependent on the physicochemical and elastic properties of the added admixture. This states that as water is used as pore fluid, the influence of the mechanical factors would remain same with an overall reduction in liquid limits of the treated soils on addition of an additives. The addition or treatment of the test soils with the crushed waste plastics not only achieved a stiff consistency material, but achieved non-frost-susceptible materials with PI less than 15. This behavior is responsible for the pavement failures resulting from the formation of frosts. With the addition of CWP and reduction of plasticity index below 15, the exercise achieved a more durable treated matrix as subgrade foundation material.

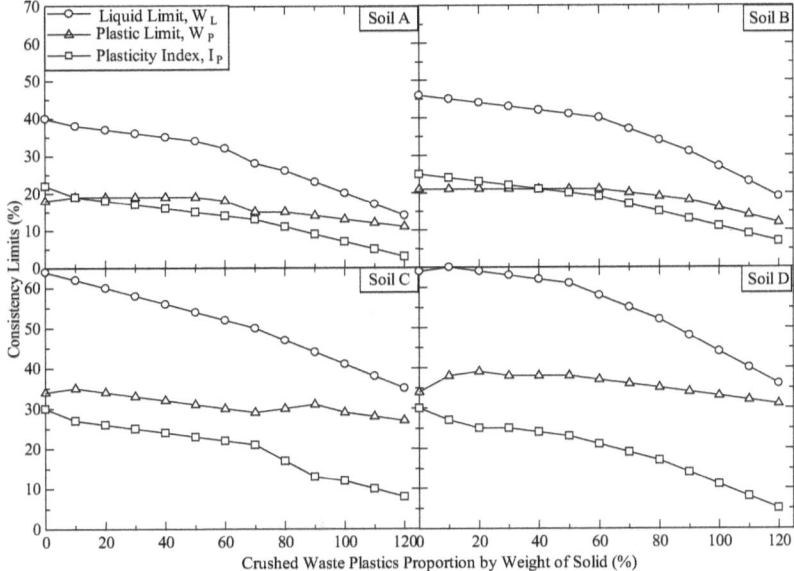

Fig. 5.1.4 Effect of CWP on the Consistency Limits of Treated Soils

5.2 Compaction

Compaction is the technique of soils re-engineering where the density of soils are improved upon through impact.

Research results have shown that the materials derived from solid wastes exhibit binding properties because of their aluminosilicates content and applied to soft soils as admixtures in a stabilization process improve the maximum dry density consistently with increased proportions of the additives. These corresponding results are due to the cation exchange reactions and the admixtures filling the voids within the soil matrix. In addition, it is due also to the formation of floccs and agglomeration of the clay particles due to polarization, release and exchange of ions. The cause of this trend is because there was increasing desire for water, which makes up with the higher amount of admixtures. This is because more water was needed for the dissociation of the ions of admixtures with Ca^{2+} and OH^- ions to supply more Ca^{2+} for the cation exchange reaction. It was also observed that optimum moisture content reduced consistently with addition of these additives. This is due to the formation of floccs within the clay complex. The additives are highly pozzolanic materials and require water for hydration thereby improving the dry density development of the treated soils. Compaction behavior of soft soils treated with additives derived from solid waste, ash and powder is a density/moisture behavioral study. This translates to the densification achieved in soils during stabilization. This is a commonly used method of soft soils treatment achieving densification by mechanical methods and exerting compactive efforts and eventually reducing the voids in treated additive/soils blends. This achieved to enable engineering soils withstand subsequent load without suffering immediate compression and this can be separated from those behavior characteristics initiated by long-term consolidation of soft clay soils. So it is very important to study and determine the moisture-density behavior of untreated and treated soils to establish improvements from the former. Ash materials are amorphous and contain ions dissociated when they react with clay minerals and water in a cation exchange reaction giving rise to ion combinations that support strengthening. The crushed materials (powder/dust) also act as fillers to improve on the porosity of the treated samples thereby improving the density. Results have shown significant reduction of the optimum moisture content (OMC) of treated soils more especially with the ash or powder based geopolymer cements because of the improved hydration reaction and moisture resistance of the polymers. The maximum dry density (MDD) obtained at OMC has also consistently improved in its properties under the influence of ash, powder and geopolymer cement additives in varying proportions. The intergranular pores are improved upon by the addition of ash, powder or ash/powder based geopolymer cement, which eventually encourages densification during compaction exercises. Shape of soil grains and amount of type of clay minerals present in treated soils are soil dependent factors that influence the behavior of soft clay soils during stabilization operations. However the addition of these additives has improved the interface behavior of these factors.

The first case under consideration here is the behavior of the maximum dry density of the crushed waste ceramics (CWC) treated soils A, B, C, and D at optimum moisture presented in Fig. 5.2.1. The proportion of additve was varied for this case between 0% and 120% of CWC by weight of the treated soils and the improvement was monitored and recorded accordingly. There was a consistent increase in the maximum dry density with a corresponding decrease in the OMC consistent with the increased proportion of the additive. This behavior was observed to be the same with all the test soils. The specific gravity responded with almost an equal increase at increased proportion of the crushed waste ceramic (CWC). The specific gravity behavior was also consistent with the test soils. The consistent increase in the MDD recorded throughout the test cycles was due to the formation of compounds of calcium and aluminum at the adsorbed complex. This characteristic was as a result of the formation of compounds responsible for strength gain through the formation floccs in the treated matrix. This behavior was also due to cation exchange reactions, flocculation, calcination reaction and the filling of the voids within the soil matrix thereby improving the porosity. And in addition, the flocculation and agglomeration of the clay minerals as a result ionization and dipolation, release and exchange of ions. The reduced moisture utilization was as a result of hydration reaction. Because moisture water was required for the dissociation of constituents with Ca^{2+} and OH^- ions to release more Ca^{2+} for the cation exchange reaction.

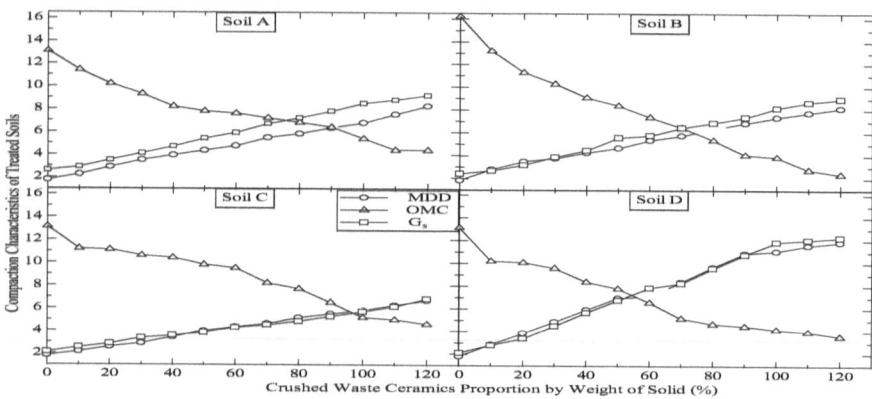

Fig. 5.2.1 Effects of CWC on Compaction Characteristics of Treated Soils

The second case is the compaction behaviour of QDbGPC to DOPC treated soil with crushed waste glasses (CWG) in a direct inverse proportion. The compaction results are presented in Fig. 5.2.2. The compaction behaviour is the densification process observed on the QDbGPC/ DOPC treated soil under the influence of added proportions of crushed waste glasses. The test soil was observed to be an unstable soil and was treated alternately with QDbGPC and DOPC in the ratios of 0:0, 0:40, 5:35, 10:30, 15:25, 20:20, 25:15, 30:10, 35:5, and 40:0% respectively. The effect of 4%, 8%, 12%, 16% and 20% by weight of crushed waste glasses over the cemented test soil was also observed. While 0:0% of the cements proportion by weight of solid served as the control point, the proportions of GPC increased from 5% in an increment of 5% while DOPC decreased from 40% at the rate of 5% also. The maximum dry density of the test soil increased with increased proportion of GPC and decreased proportion of DOPC. This consistently continued until 40:0% corresponding to GPC and DOPC respectively. The specific gravity also increased in that succession consistently. Alternatively, the optimum moisture content decreased with the same pattern. Notably, the introduction of the high content aluminosilicate crushed waste glasses improved the compaction characteristics of the test soil under the influence of the cements. This behaviour on the axes of the cements linear inverse replacement process was due to the introduction of a more bio-based cementing material, which is resistant to sulphate attacks, cracking and brittleness. Also, the bio-based cementing geomaterial i.e. the QDbGPC produced more silica and aluminate to form CSH and CAH responsible for strength gain and densification. It forms more elastic agglomeration and sequestrum and floccs to produce a more densified treated soil. Cation exchange reactions between the dissociated ions from the bio-based cementing material caused the increased density and specific gravity with increased proportions of GPC. These increased MDD were obtained at optimum moisture content.

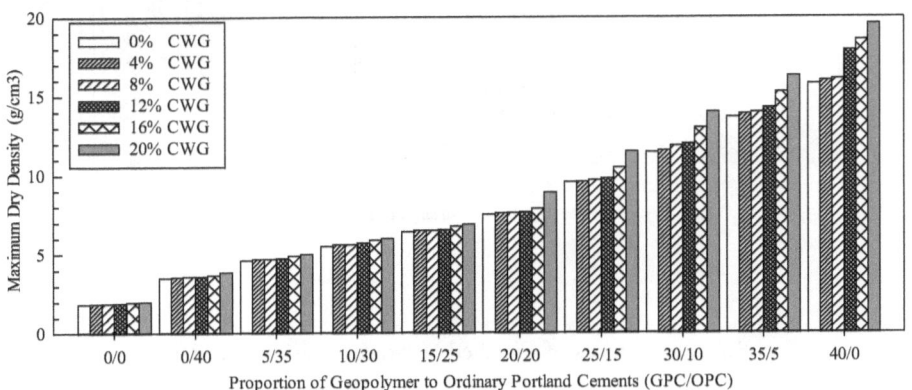

(a)

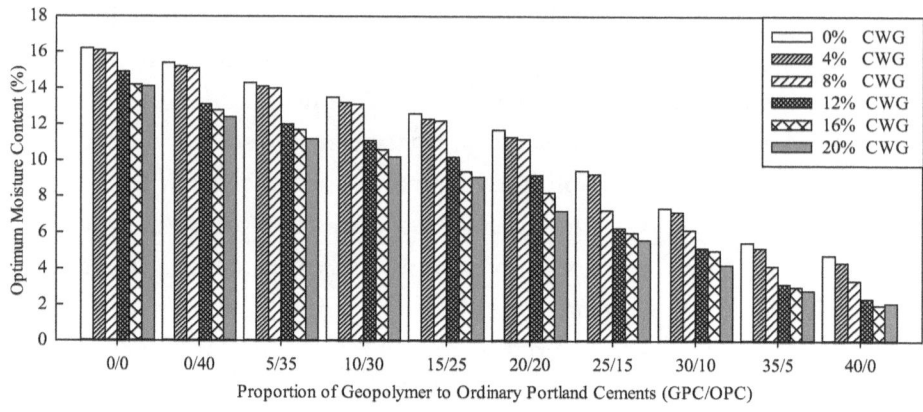

Fig. 5.2.2 Influences of Crushed Waste Glasses on Compaction behaviour of treated soil: (a) Maximum Dry Density, (b) Optimum Moisture Content, (c) Specific Gravity

The third case is the compaction behaviour of QDbGPC treated uncemented lateritic soils. Similarly, the results of the compaction behavior of the treated lateritic soils are presented in Fig. 5.2.3. The maximum dry density (MDD) improved consistently from 1.76, 1.85 and 1.80 g/cm³ respectively at 0% by weight utilization of quarry dust based geopolymer cement to 8.5, 10, and 7.8 g/cm³ respectively at 150% by weight utilization of quarry dust based geopolymer cement. On the other hand, the optimum moisture reduced in a consistent trend also. Moisture was required to cause a dissociation of cations and anions from the geopolymer cement during hydration to supply more Ca^{2+} for the cation exchange reaction. This cation exchange reaction caused the consistent reduction of moisture content, which led to the formation of floccs of the clay particles hence the densification of the treated soils.

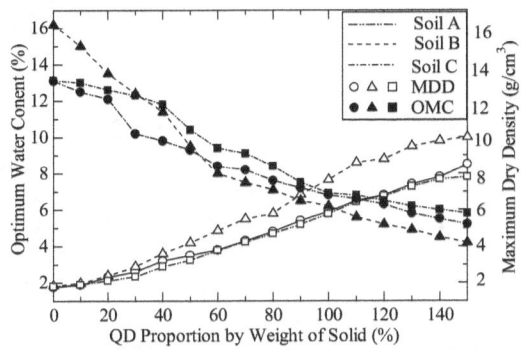

Fig. 5.2.3 Compaction Behaviour of QDbGPC Treated Soils

The fourth case is the behavior of the maximum dry density of the crushed waste plastic (CWP) treated soils A, B, C, and D at optimum moisture have been presented in Fig. 5.2.4. The proportion of additive varied between 0% and 120% of CWP by weight of the treated matrix. There was a consistent increase in the maximum dry density with a corresponding decrease in the OMC consistent with the increased proportion of the additive. This behavior was observed to be the same with all the test soils. The specific gravity responded with almost an equal increase at increased proportion of the crushed waste ceramic (CWP). This specific gravity behavior was also consistent with the test soils. The consistent increase in the MDD recorded throughout the test cycles was due to the formation of compounds of calcium and aluminum at the adsorbed complex. This characteristic was as a result of the formation of compounds responsible for strength gain through the formation floccs in the treated matrix. This behavior was also due to cation exchange reactions, flocculation, calcination reaction and the filling of the voids within the soil matrix thereby improving the porosity. And in addition, the flocculation and agglomeration of the clay minerals as a result ionization and dipolation, release and exchange of ions. The reduced moisture utilization was as a result of hydration reaction. Because moisture water was required for the dissociation of constituents with Ca^{2+} and OH^- ions to release more Ca^{2+} for the cation exchange reaction.

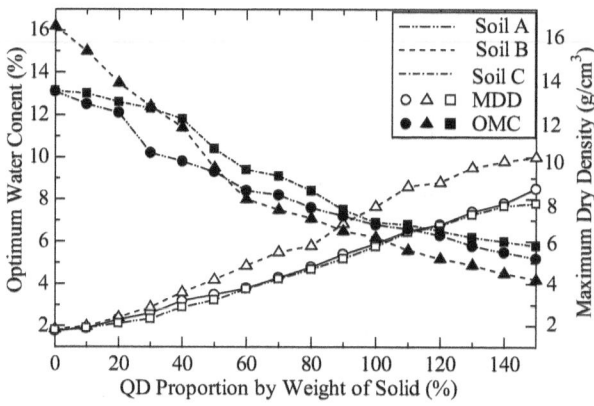

Fig. 5.2.4 Effects of CWP on Compaction Characteristics of Treated Soils

The fifth case is actually a specialized approach that studied the behavior of the re-engineered soils under the influence of quarry dust based geopolymer cement and varying compactive efforts. The Geopolymer cement treated test soil A which was cemented with 5% Portland cement showed a consistent strength density gain at zero addition of Geopolymer cement under varying compaction blows and compactive efforts. The use of the modified proctor mold rammer of 4.5kg produce a higher compactive energy than the standard proctor mold rammer of 2.5kg. Again, the blows between 6 and 55 produce different density gain which increased with increased blows. Subsequently, the quarry dust based Geopolymer cement was added in the proportion of 2.5%, 5%, 7.5% to 20% in that order which equally produced significant strength gain. It was so remarkable that a control density of between 1.78 and 1.97g/cm³ at zero Geopolymer cement produced densities in the range of 3.89 to 6.89g/cm³ at 20% by weight addition of quarry dust based Geopolymer cement under compaction blows of between 6 and 55 as presented in Fig.5.2.4. It is important to note that this behaviour may be due to the increase compactive effort produced by both rammer sizes and number of blows per layer, which enhanced the interlocking between the microstructure of the soil and in turn reduced the porosity in the treated matrix, thereby increasing densification and flocculation, consequently higher strength might be gained. This point to the fact that higher compaction effort may encourage the use of little or no Portland cement reducing cost, and shrinkage potential. Secondly, the increased proportions of quarry dust based Geopolymer cement may have caused a consistent strength gain in the treated matrix. This behaviour gained is because the Geopolymer cement may have filled the voids within the soil mass during the stabilization procedure to improve the porosity of the treated soils. Additionally, because of its pozzolanic properties which enhanced calcinations reaction, pozzolanic reaction, and the inclusion of Na_2SiO_3 in NaOH solution, Geopolymer cement provides higher silicate concentration and gives rise to the formation of gel which likely fastened polymerization, and consequently polycondensation which led to the obvious gain in strength of the treated soils. The Na_2SiO_3 acted as a nucleating site then increased with the amount of silicates released, leading to the formation of more hydration points. As the concentration of hydration materials increased, the number of contact points between hydration materials also increased, consequently forming a solid microstructure within the treated soils matrixes. As presented in Fig. 5.2.5 and 5.2.6, the proportions of Portland cement were reduced to 2.5% and 1%, respectively. These percentages were used to cement the test soils at varying proportions of quarry dust based Geopolymer cement. The behaviour of the treated soils showed almost the same results, but with a very little reduced strength compared to the behaviour with 5% Portland cement cementation. At 0% Portland cement as shown in Fig.5.2.7, the treated soil A gained significant strength because it recorded between a range of 1.74 and 1.95g/cm^3 at zero Geopolymer cement (Fig. 5.2.7a, c), and of 3.37 and 5.89 g/cm^3 at 20% by weight Geopolymer cement under varying compactive efforts, which were almost the range strength density gain at 5, 2.5 and 1% addition of Portland cement. The optimum moisture content (OMC) at which these densities were recorded was observed to decrease consistently too over the increased compaction blows and increased proportions of quarry dust based Geopolymer cement (Fig. 5.2.7b, c). This may be due to the finely ground quarry dust based Geopolymer cement which acted as a filler material, leading to decreased porosity and then reduced moisture at which maximum densification was achieved. This behaviour shows that quarry dust based Geopolymer cement possesses the properties with which to completely replace ordinary Portland cement in the stabilization protocol, which is a positive turn as it bothers on environmental effects of cement usage in geotechnical engineering operations and civil engineering works as a whole. Previous study found that for one ton of ordinary Portland cement produced or used, an equivalent one tonne of CO_2 emission may be released into the atmosphere, which contribute to global warming and this is an environmental global issue.

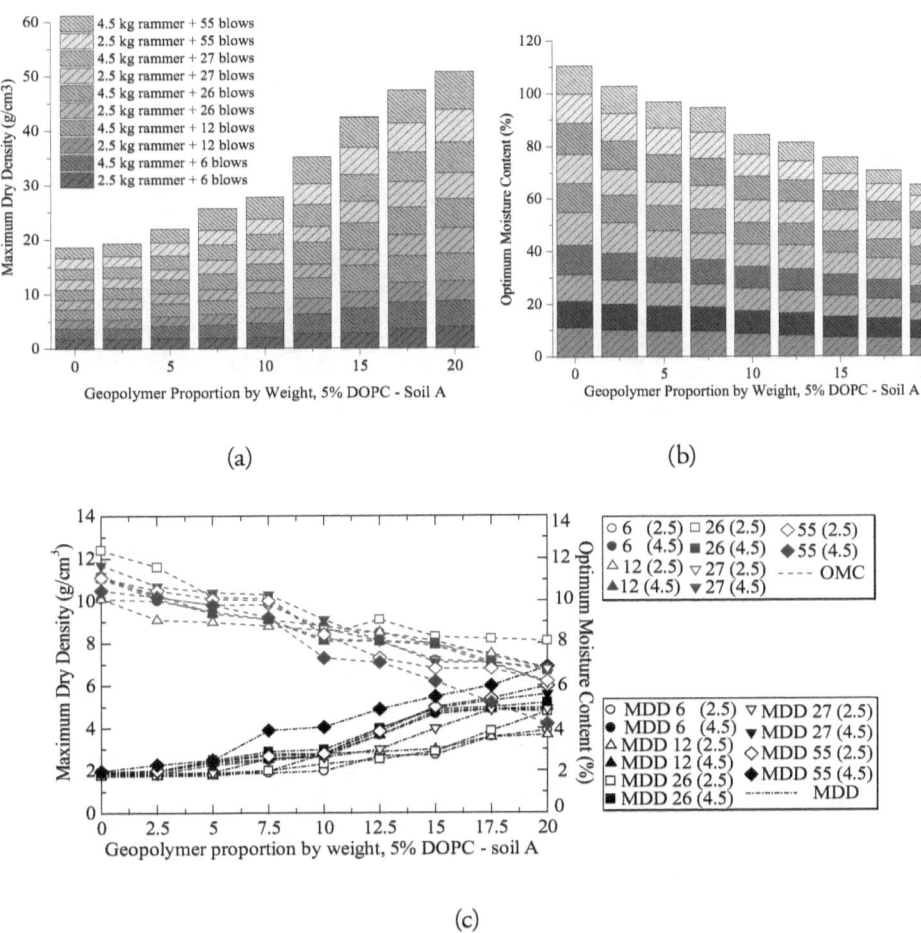

Fig. 5.2.5. Compaction of Test Soil A at 5% DOPC and Varying Proportions of Quarry Dust based Geopolymer under Varying Blows

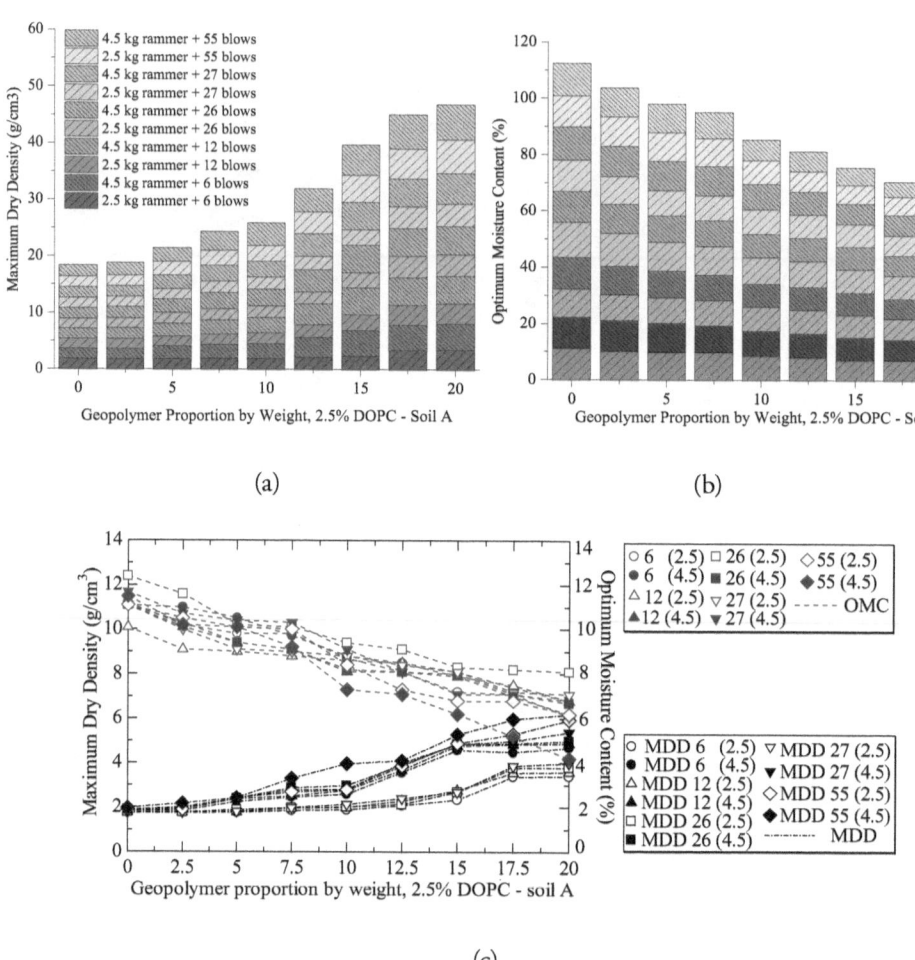

Fig. 5.2.5. Compaction of Test Soil A at 2.5% DOPC and Varying Proportions of Quarry Dust based Geopolymer under Varying Blows

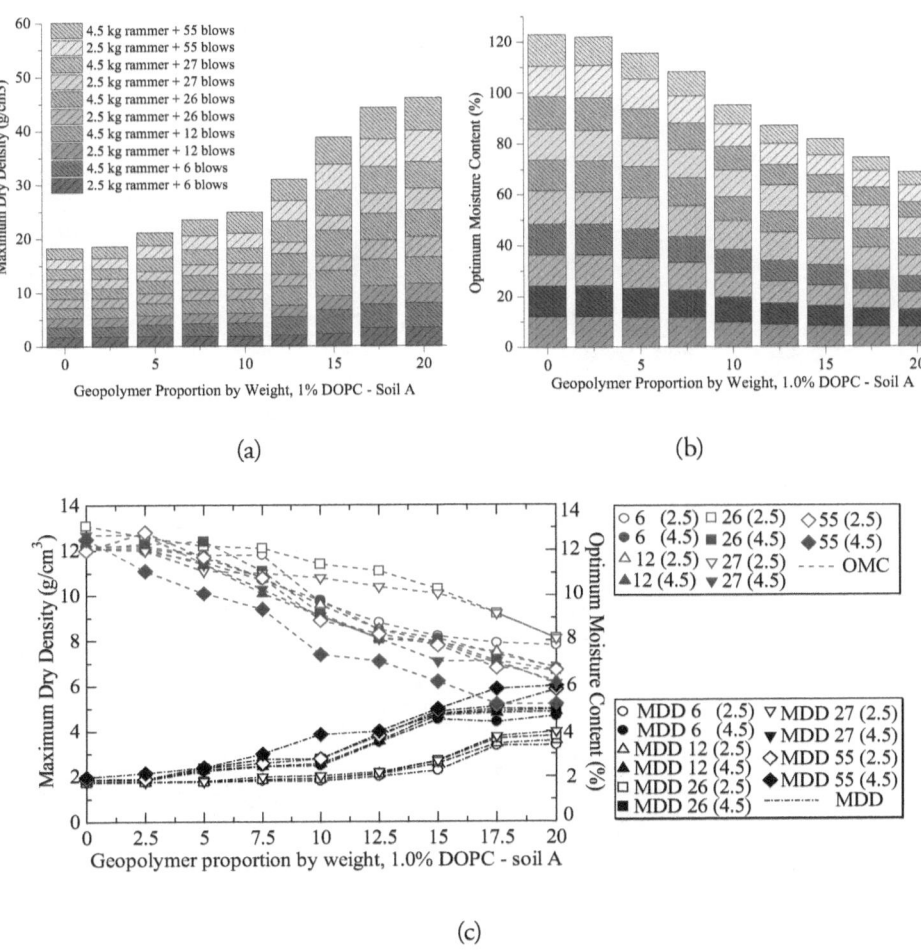

Fig. 5.2.6. Compaction of Test Soil A at 1% DOPC and Varying Proportions of Quarry Dust based Geopolymer under Varying Blows

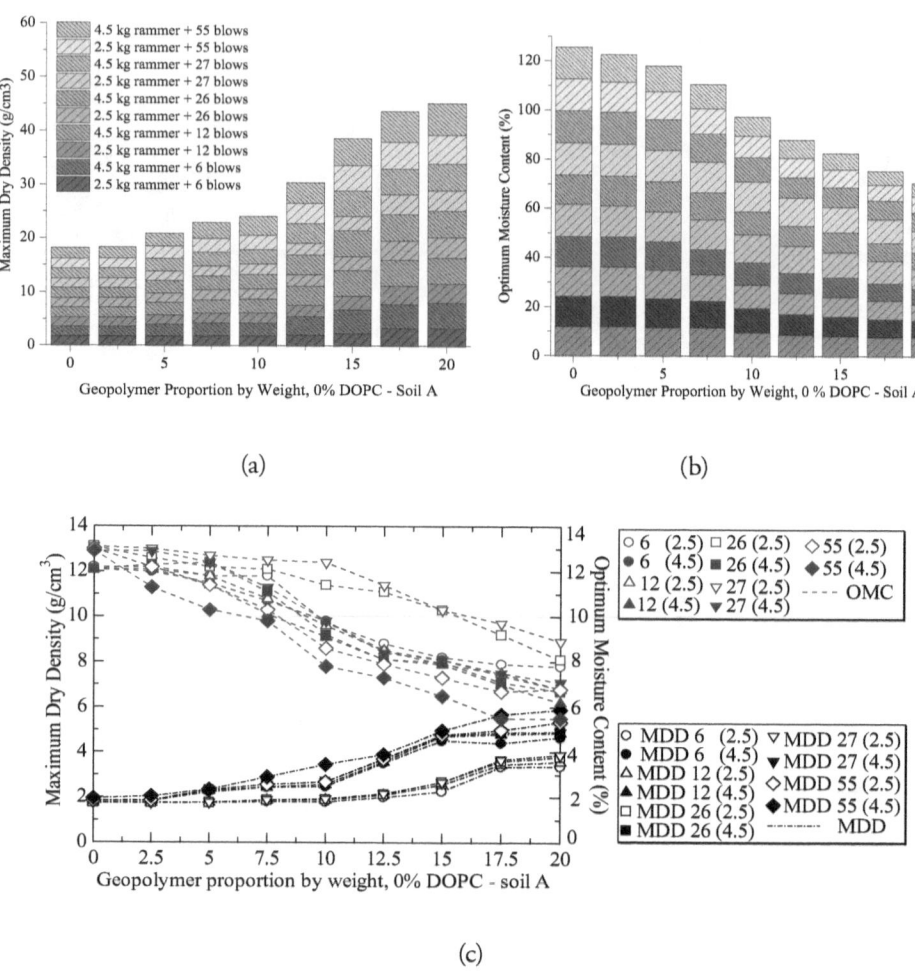

Fig. 5.2.7. Compaction of Test Soil A at 0% DOPC and Varying Proportions of Quarry Dust based Geopolymer under Varying Blow

A further illustration was carried with soil B to establish the changes in behavior of soils re-engineered in similar pattern but under varying campaction. The compaction characteristics of soil B at 5% Portland cement were displayed in the Fig.5.2.8a, 5.2.8b, and 5.2.8c, respectively. The obtained results indicate an analogous behaviour of treated soil B to that found in soil A, that is having a steady strength density achieved at zero addition of Geopolymer cement under varying compaction blows and compactive efforts. In addition, the use of the higher hammer weight of 4.5 kg along with the longer drop distance involved in the modified proctor mold brought about a greater compaction energy compared to that in standard proctor mold rammer of 2.5kg. Analogously, different number of blows of 6, 12, 26, 27, and 55 resulted in different density gain, which increased with increased blows. The influences of various proportions of the quarry dust based Geopolymer cement on compaction behaviour of treated soil B were also examined by adding proportions of Geopolymer cement of 2.5, 5, 7.5 to 20% in that order. Generally, the experimental results show a consistent strength density gained as higher ratios of Geopolymer cement were added into the study soil B. Specifically, a control density of between 1.77 and 3.12 g/cm^3 at zero Geopolymer cement produced densities in the range of 3.38 to 5.89 g/cm^3 at 20% by weight addition of quarry dust base Geopolymer cement under compaction blows of between 6 and 55 as presented in Fig.5.2.8 a, c. The increase in density of treated soil B may firstly be attributed to the increase in applied compaction energies which originated by both hammer weight and numbers of blows per compaction layer. Due to these two weighty factors, the voids between the treated matrixes could be significantly diminished, which in turns increase densification and flocculation of treated soil mass, consequently a higher strength obtain.

Additionally, since the quarry dust based Geopolymer cement is added into the soil B, the voids within the soil mass tend to be filled during the stabilization procedure. Due to pozzolanic properties of the Geopolymer cement which enhances cation exchange reaction, pozzolanic reaction, and polycondensation, the presence of the Geopolymer cement could contribute to the development of density as well as strength of treated soil B.

The effects of ordinary Portland cement on the compaction characteristics of treated soil B were also considered. In Fig. 5.2.9 and 5.2.10, the proportions of Portland cement were lowered to 2.5% and 1%, respectively. The aim of adding these percentages of Portland cement was to bind the test soil at varying proportions of quarry dust based Geopolymer cement. Similar to that found in the treated soil A, compaction behaviour of treated soil B was found almost the same, with a trivial reduced density compared to that was gained at 5% Portland cement cementation. At 0% Portland cement as presented in Fig. 5.2.11, the treated soil B gained significant strength because it recorded between 1.74 and 2.95g/cm^3 at zero Geopolymer cement and 3.37 and 5.69g/cm^3 at 20% by weight Geopolymer cement under varying compactive efforts, which were almost the same range of strength density gain at 5, 2.5 and 1% addition of Portland cement. This experiment research on compaction characteristics of the test soil B treated with quarry dust based Geopolymer cement proving that the quarry dust based Geopolymer cement exhibits the properties which able to partly or even fully take the place of Portland cement in the soil stabilization goal. The optimum moisture content at which these densities were recorded was observed to decrease consistently too over the increased compaction blows and increased proportions of quarry dust based Geopolymer cement.

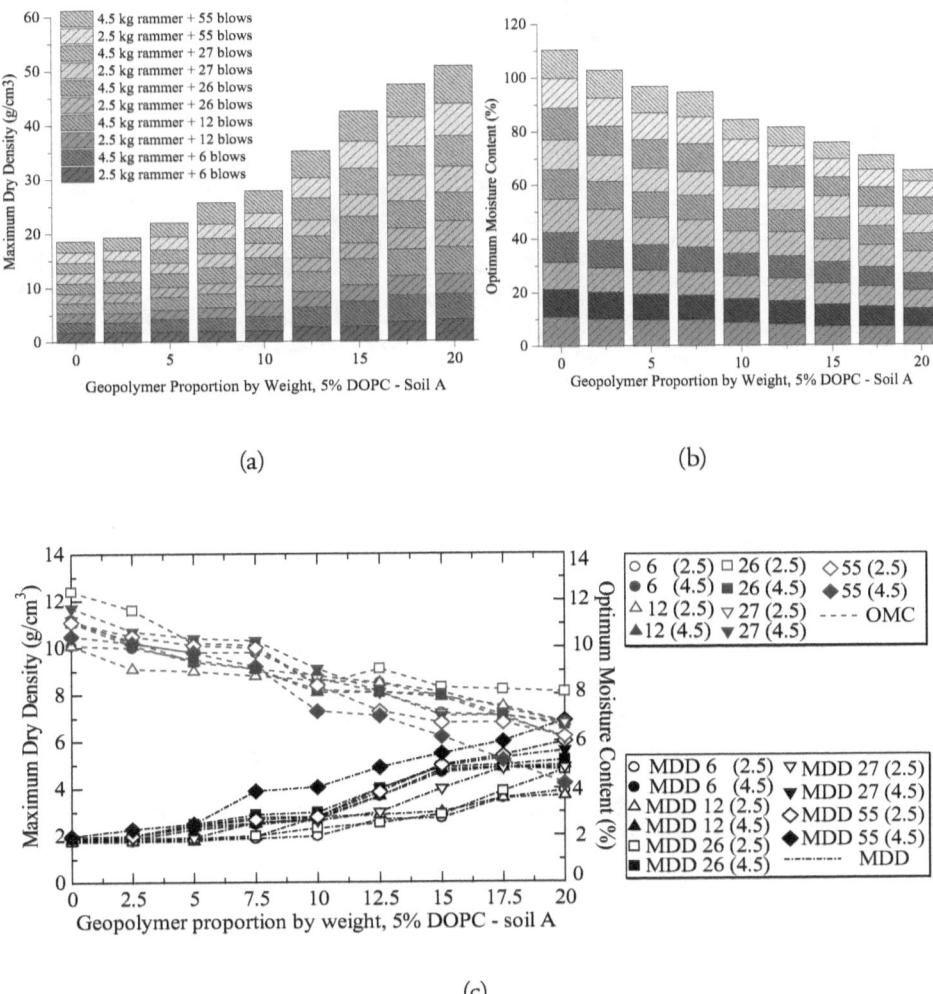

Fig. 5.2.8. Compaction of Test Soil B at 5% DOPC and Varying Proportions of Quarry Dust based Geopolymer under Varying Blows

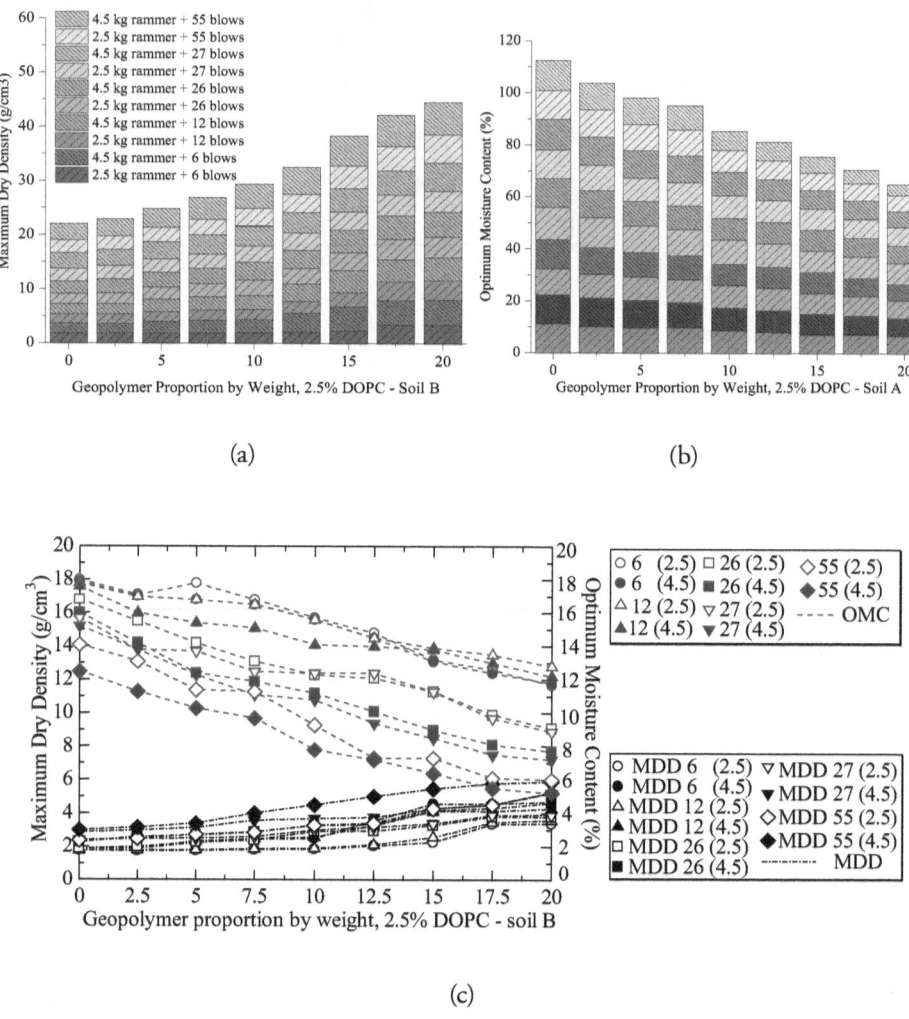

Fig. 5.2.9. Compaction of Test Soil B at 2.5% DOPC and Varying Proportions of Quarry Dust based Geopolymer under Varying Blows

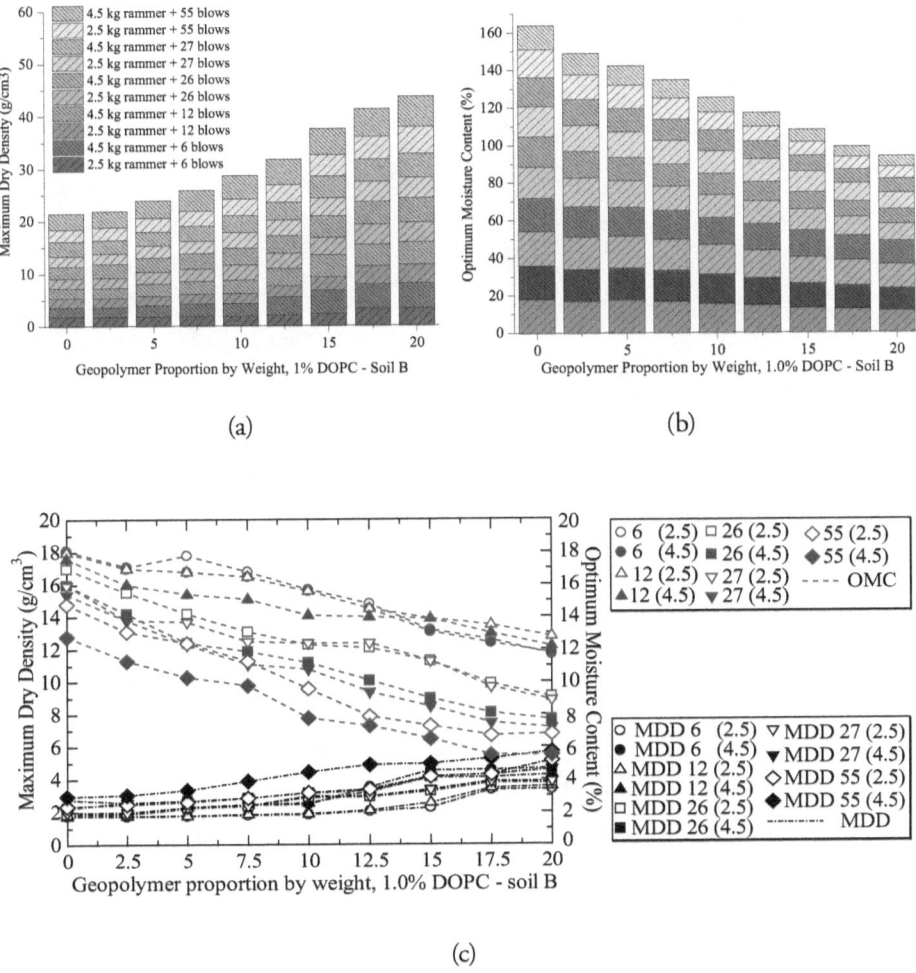

Fig. 5.2.10. Compaction of Test Soil B at 1% DOPC and Varying Proportions of Quarry Dust based Geopolymer under Varying Blows

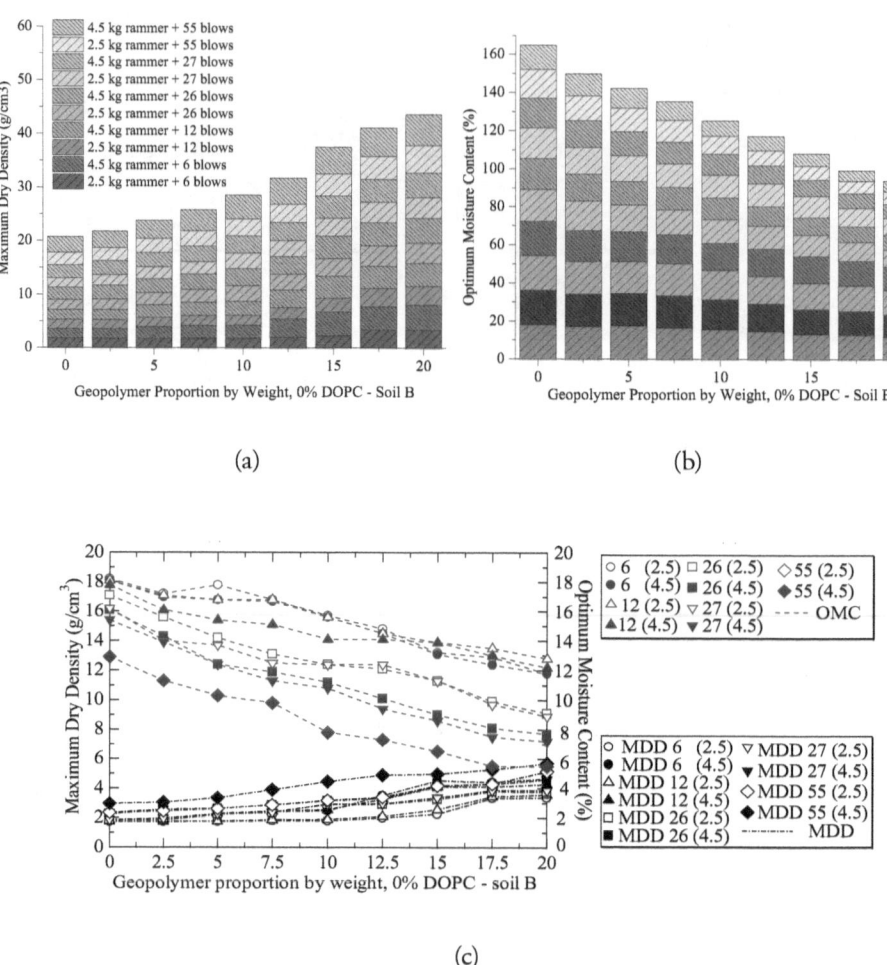

Figure 10. Compaction of Test Soil B at 0% DOPC and Varying Proportions of Quarry Dust based Geopolymer under Varying Blows

5.3 Volumetric Changes

Swelling and shrinkage properties are very important factors that influence the behavior of engineering soils as foundation materials especially in hydraulically bound environments. Soils as foundation materials suffer moisture attacks through capillary action, suction, rise and fall of water table and lateral percolation. This is more prevalent in pavement structures. Volume changes cause undesirable characteristics that impair the performance and long-term service life of foundations and pavement infrastructures. Experimental results have shown that these undesirable properties of soft clay soils have been arrested by the addition of ash or powder or ash/powder based geopolymer cements in a stabilization procedure. First, the moisture resistance ability of the ash or powder materials derived from solid waste due to their composition and the hydration structure makes it possible for soils treated with them to behave within acceptable swelling limits. There have been recorded reduction in swelling potentials through the addition of these additives. Conversely, these materials possess the ability to resist drying during the reverse phase of swelling i.e. reduction of entrapped moisture. Shrinkage in soft clay soils leads to cracking and lateral deformations but the additions of solid waste derivatives has changed this undesirable behavior due to the ionic composition of aluminosilicates admixtures.

The first situation in Fig. 5.3.1 has presented the graphical behavior of the swelling potential of the QDbGPC treated test lateritic soils. From the reference experiment with the untreated natural soil, an increased swelling potential at increased curing time was recorded. This exhibited an exponential rise in vertical swelling between 56 and 70 days of curing which represented a swell rise rate of 35% compared an average swell rise rate of 9% at and between 0-56 days curing. On the other hand, the QDbGPC treated soils behaved different with the increase in the geopolymer cement utilization. The swelling potential improved with increase in the utilization of the quarry dust based geopolymer. More of this improvement was observed between 30% and 120% by weight utilization of the geopolymer cement. This is because of the fact that water dipoles were absorbed more between platelets under a condition attributed as microscopic swelling at and between 30% and 120% by weight utilization of the geopolymer cement. The high content of sodium silicates activator due to increased quarry dust based geopolymer tended to increase the release of Al^{+++}, Si^{++++}, and Ca^{++} from the grains of metallurgical slag. These cations released in turn raised the geopolymerization and calcinations reactions at the diffused layer of the test soils and additive materials interface. More hydration points were formed due to further release of silicates from the Na_2SiO_3, which acted as a surface of nucleation. The reaction interface between mixed matrixes of test soils also increased as the potential of materials of hydration increased however forming a microstructure of solid consistency. The resultant outcome of this reaction is the reduction in swelling tendency of the treated test soils. The porosity of the treated soil may have been improved by the addition of the quarry dust based geopolymer cement due to its ability to fill in voids and to form floccs. Conversely, the property that represents the ability of a treated or untreated matrix of soil to withstand crack tendencies was examined as drying shrinkage and it's presented in Fig. 5.3.2. The three test lateritic soils were treated with QDbGPC at the same proportion by weight utilization. Treated and untreated test soils change in volume when a reduced shrinkage limit is observed. So, increased values of shrinkage limits presents an improvement on the shrinkage behavior of the treated soils as can be deduced from the graphs representing test soils A, B, and C. The untreated natural soils A, B and C recorded shrinkage of 8%, 7% and 7% respectively. It is also recorded that the shrinkage limit increased with increase in both quarry dust based geopolymer utilization and drying time. At 4 hours oven drying time, the average improvement rate of 9% was recorded while beyond 4 hours oven drying time, the improvement rate increase from 15% to 23% with increased drying time. Similarly, the improvement increased with increased in the QDbGPC utilization in the treatment of the test soils. This behavior was due to the ability of geomaterials to resist shrinkage tendencies and prevent cracking effects. The increased quantity of geopolymer cement utilization in the treatment protocol released high concentration of silicates at the diffused surface of the clay reaction interface forming an elastoplastic band of matrix. The matrix has the ability to resist shrinkage effects resulting from medium moisture and atmospheric variations. Consequently, these materials could best fit into a moisture bound environment like the pavement foundation environment thereby resisting moisture variation effects that lead to swelling and shrinkage.

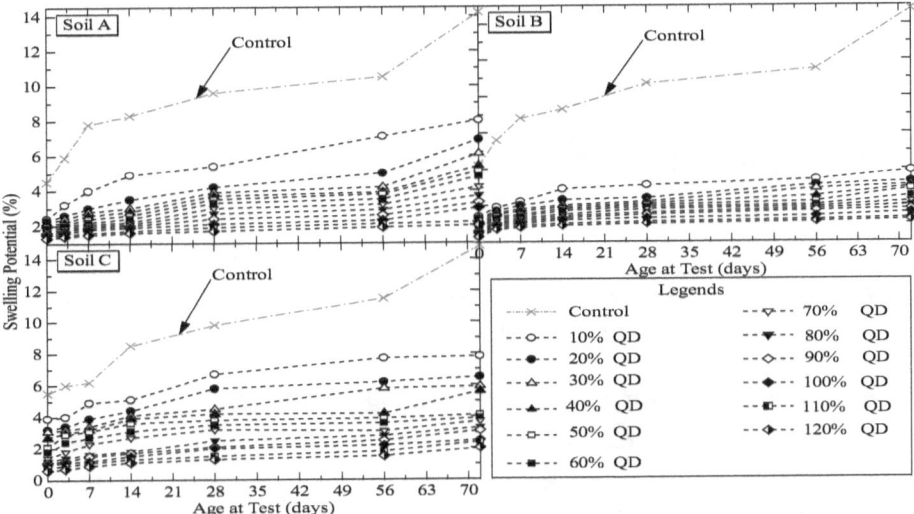

Fig. 5.3.1 Effect of Quarry Dust Propotion on Swelling Potential Behaviour of Treated Test Soils: Soil A, Soil B, and Soil C

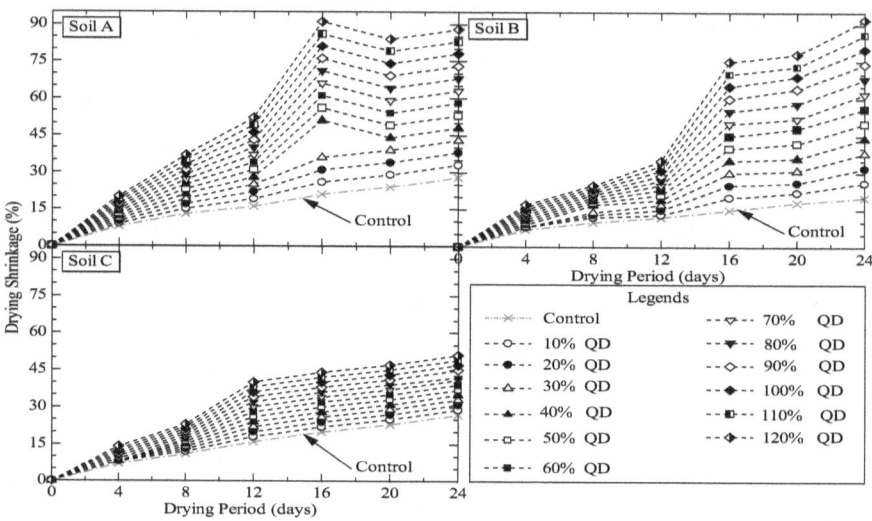

Fig. 5.3.2 Effect of Quarry Dust Propotion on Shrinkage Behaviour of Treated Test Soils: Soil A, Soil B, and Soil C

The second situation is presented in Fig. 5.3.3 which presents the swelling potential behaviour of the treated soils at the same rate and immersed at varying number of hours taken precisely at 0, 3, 7, 14, 28, 56 and 72 hours. Results have shown that the addition of quarry dust at the same rate in an increment of 10% consistently reduced the swelling potential of the treated soils. Though the swelling potential increased at increased duration of immersion in moisture, but the improvement recorded at the addition of the quarry dust was remarkable that the increase in swelling with soaking time becomes insignificant. This is due to the hydration rate of the quarry dust additive during the hydration and carbonation reactions that formed floccs of the QD treated soil blend reducing the swelling effect on the soils. On the reverse cycle presented in Fig. 5.3.4 i.e. shrinkage, the shrinkage limit improved considerably with increased quarry dust addition. Thus is due to the carbonation reaction between the soil dissociated ions and the eco-friendly additive with the characteristic feature of improving heat, crack and brittle resistance of materials treated with alternative cementing materials (binders).

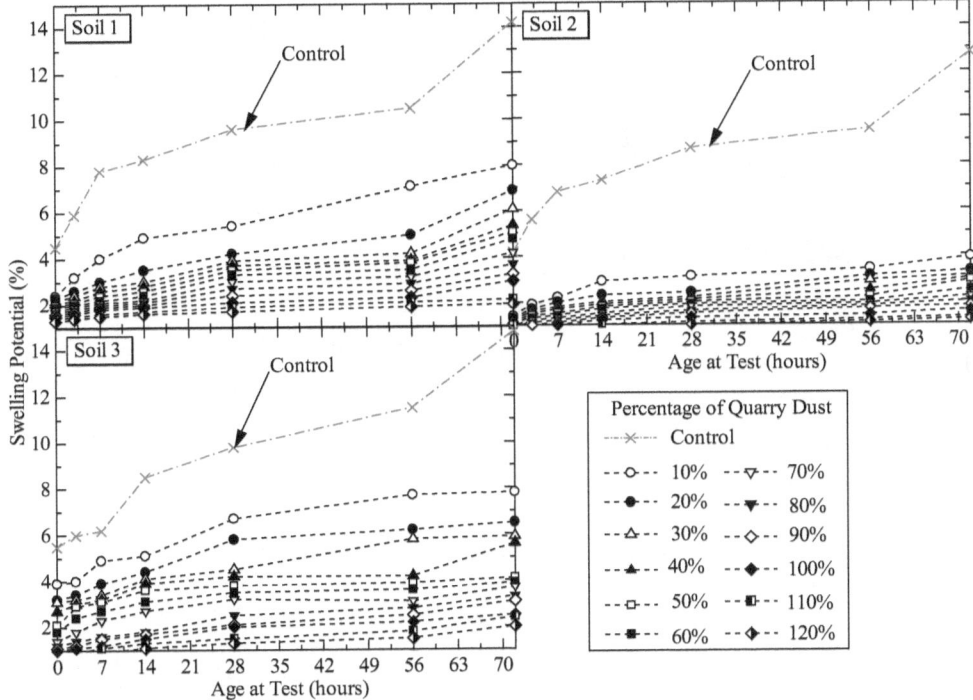

Fig. 5.3.3 Effect of Quarry Dust Propotion on Swelling Potential Behaviour of Treated Test Soils: A, B, and C

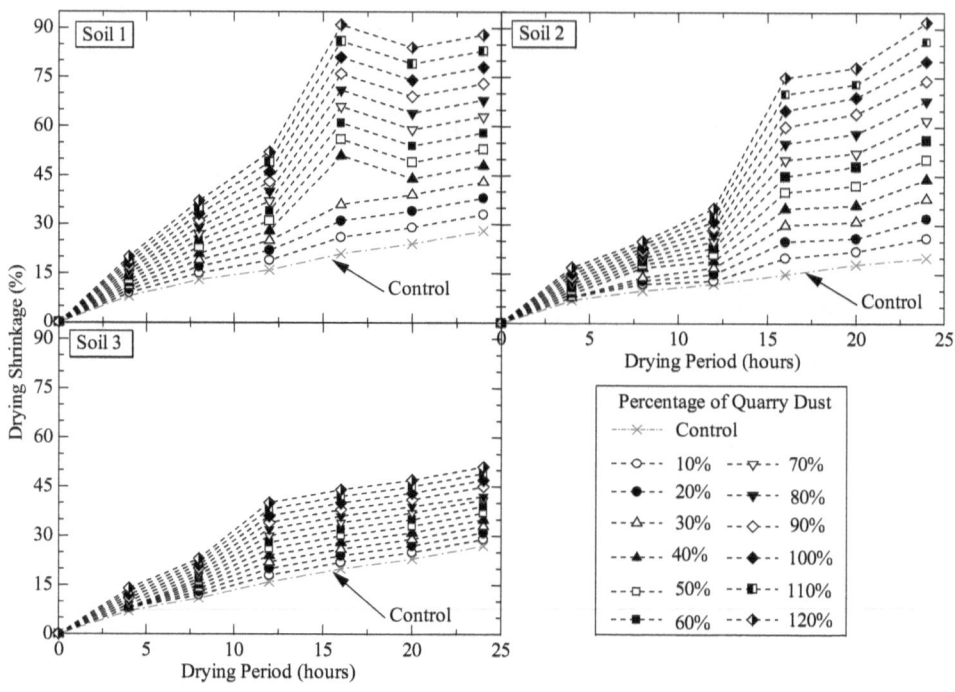

Fig. 5.3.4 Effect of Quarry Dust Propotion on Shrinkage Behaviour of Treated Test Soils: A, B, and C

5.4 Sorptivity

A pavement section may generally be defined as the structural material placed over a subgrade layer. In asphaltic pavement it is typically a multi-layered system comprising the subgrade (support), the subbase, base course and surfacing. Its principal function is to receive load from traffic and transmit it through its layers to the subgrade. These pavements are to aid our safe movements alongside the goods being transported on it. This fact has faced with much challenges and so much has being done to keep making better the already provided remedies. Among the serious factors impending the sustainability of our pavements is a bad subgrade or foundation, amongst which are expansive soils. The clayey soil has the capacity to expand at the increase of its moisture content likewise shrink when its moisture content decreases due to the presence of Montmorillonite clay mineral. These soils are known as Black cotton soil or Expansive soil. The moisture may come from rain, leakage from water or sewer lines, seepage from nearby water bodies, earthen drains, variation of ground water table. These soils have high plasticity, high shrinking and swelling characteristics, extremely low shear strength, bearing capacity and high compressibility. These soils are considered as poor soil regarding the engineering purposes. Soil stabilization is the best-known method of making use of clays and can decrease the volume change of such clay due to change in water with the help of stabilizers. Such stabilizers are mixed with the soil to reduce the swelling potential and plasticity of expansive clay, improving the durability and strength of the soil. Hence, an attempt has been made to understand the potential of quarry dust cushion to prevent seasonal swell–shrink in expansive soil. Stone dust is a kind of solid waste material which is generated from the crusher industries. It is estimated that a crusher unit produces 15% to 20% stone dust of its total production of stone aggregates. It possesses a great problem of disposal due to the limited land area and environmental issues. The best way to eliminate this problem is to make use of this waste. Stone dust also referred to as quarry dust is available at nominal cost in almost all the places. One of the aim of stabilizing with cementing additives such as quarry dust is to improve physical–mechanical properties of the expansive soil amongst which is Porosity. Soil porosity refers to the amount of pore or open space between soil particles. Due to problems associated with porosity are studied by the absorption test and permeability tests. It becomes expedient to mention that permeability which is a measure of the flow of water under pressure in a saturated porous medium does not incorporate the measure of the rate of absorption by capillary suction. This warranted a need for another type of test. This test should measure the rate of absorption of water by capillary suction. Another lore to be pointed out at this point is sorptivity, which characterizes the material's ability to absorb and transmit water through it by capillary suction. Whilst permeability is an important parameter for water retaining structures, a more important parameter (which is directly related to durability) for above ground structures is sorptivity. Rushabh in 2013 mentioned that sorptivity of a material is the ability to absorb and transmit water through it by capillary suction, that it is a simple parameter to determine and is increasingly being used as a measure of a solid's resistance to exposure in aggressive environments and same can be said of clay. Another monumental issue to analyse is the fact that not much has been done to investigate the properties of treated soils such as treated soft soils, in this case investigating the sorptivity, swelling, shrinkage, compression and durability of quarry dust treated soft soils for moisture bound pavement. It becomes essential to acknowledge the fact that, for sustainability to be attained for pavements that its subgrade is a treated soft soil, like a quarry dust treated clay those outlined properties of the treated soil should be understood and harnessed to make durable the pavement. When subgrades get exposed to water, at first the moisture is absorbed by capillary potential of the soil matrix. The initial water infiltration process is governed by the capillary force, after which the gravitational force takes over. The infiltration that is driven by the capillary forces alone is related to sorptivity. Cumulative absorption or desorption of water into or out of a horizontal (minimal gravity effects) column of soil with uniform properties and moisture content is proportional to the square root of time and has been termed sorptivity ($S = [LT^{-1/2}]$) by Philip. Another term associated to this is intrinsic sorptivity ($\zeta\,[L^{-1/2}]$), Philip [31] went on to define intrinsic sorptivity from the Sorptivity and fluid properties, as ($\zeta = (\mu/\sigma)S$). Where μ is the dynamic viscosity ($ML^{-1}T^{-1}$) and σ as the surface tension of the fluid (MT^{-2}). Sorptivity depends on initial uniform water content (θ_n) or potential (ψ_n) of the soil and the water content (θ_n) or potential (ψ_n) on the intake surface, so that strictly we should write sorptivity as S (θ_n, θ_0) or S (ψ_n, ψ_0). The latter form is needed when the potential at the intake surface is positive. Sorptivity can be defined analytically as a function of soil water content and diffusivity. The calculation of the sorptivity involves iterative numerical procedures, and because of technical difficulties, several approximations have been proposed. The approximation by Parlange proposed in Equation 5.4.1 has been found to give good results:

$$S_0^2 = 2\sqrt{\theta_n - \theta_0} \int_{\theta_0}^{\theta_n} \sqrt{\theta_n - \theta_0}\, D(\theta)d\theta \qquad (5.4.1)$$

Where θ_0 is the water content at applied potential, ψ_n, θ_n is the initial water content of the soil, and D is the soil diffusivity.

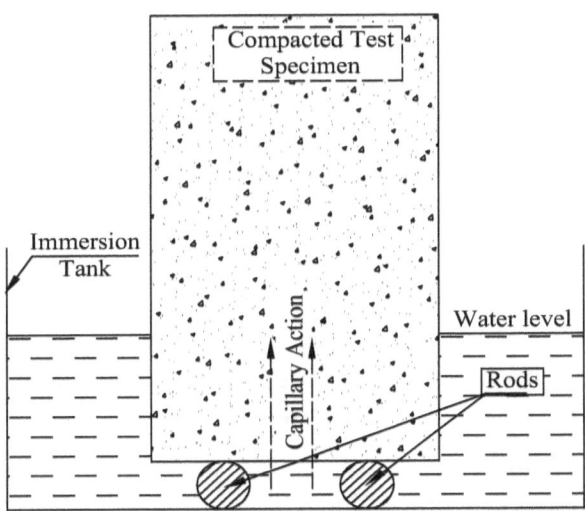

Fig. 5.4.1 Schematic Arrangement of the Sorptivity Test

Sorptivity was introduced as a testing method that consisted of a uniform directional water absorption front within a specimen. The cumulative infiltration or absorbed volume of moisture per unit area of inflow surface, I (mm) was related to the square root of the elapsed immersion time, t ($min^{1/2}$) and the relationship in Equation 5.4.2 was developed.

$$S = \frac{I}{\sqrt{t}} \qquad (5.4.2)$$

Where S is the Sorptivity ($mm.min^{-1/2}$)

Figs. 5.4.2, 5.4.3 and 5.4.4 presents the sorptivity and cumulative infiltration (suction) behaviour of the treated test soils over certain moisture exposure durations for the test soils A, B and C. This laboratory conditions through practice and research experience has been extrapolated to reflect the field conditions and can be adapted to any part of the world where soils with similar charateristics prevail. The test soils were treated with QD at the rates of 10%, 20%, 30% and 40% by weight and the observed under the laboratory conditions. The respective sorptivity behaviour as presented in Figs. 5.4.2a, 5.4.3a and 5.4.4a show a quantum line from control to a point called the nick point, which was followed by a gradual response to the immersion time. The location at which this shift in slope occurred was called the nick point because it indicated a point of saturation shift with time of exposure or immersion also indicative that at the initial phase of the moisture exposure, there was a serious capillary absorption. This behaviour gave two distinct slopes in the graphical behaviour of the treated soils sorptivity. The two distinct slopes present in the absorption curves represent the initial i.e. new-age absorption and secondary i.e. the old-age absorption. The new age gradient was typically steeper than the old age gradient, signifying the greater rate of absorption during the early periods of exposure or immersion. After an elapsed time, the change in gradient of the absorption or suction curve into the old-age absorption or suction signified the saturation of the specimens. This shows evidently and technically that the saturation of the treated specimen caused a remarkable reduction in the capillary suction of the specimens thus decreasing the rate of absorption to nearly zero at times. Due to the increased rate of sorptivity, the nick point of the specimens can occur at a very early time. The importance of the nick point should not be ignored because it indicates the degree of saturation of the treated soils specimen. This parameter suggests that the nick point can be a fundamental factor or condition when predicting or modelling the service life of treated soils used as pavement foundations or subgrade layer materials. It was observed that the secondary absorption maintained a remarkably lower rate than the initial sorptivity. Conversely, Figs. 5.4.2b, 5.4.3b and 5.4.4b present specifically the behaviour of the cumulative infiltration over varying immersion duration. This showed a consistent behaviour with exposure time and reduced consistently with increased proportion of QD addition. This behaviour is fundamentally indicative that with reduced sorptivity with increased QD, there is a consequent check on swelling potential and shrinkage limits because of the moisture relation point between these characteristics of the treated pavement material. Durability index is also a factor of moisture exposure in loss of strength on immersion technique. It is only sorptivity (suction rate) of the treated pavement materials that makes this test feasible and determinable.

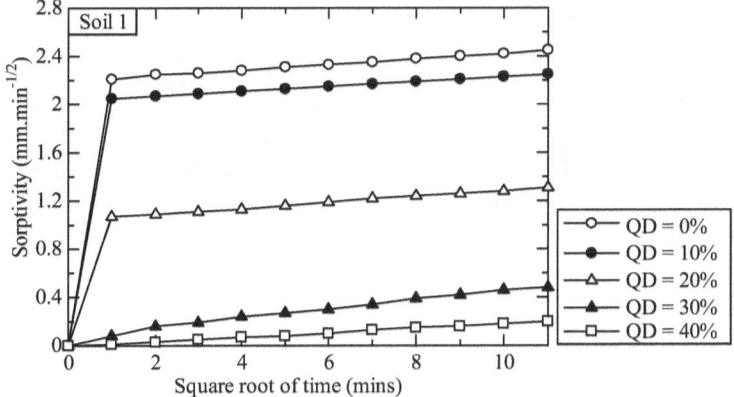

(a)

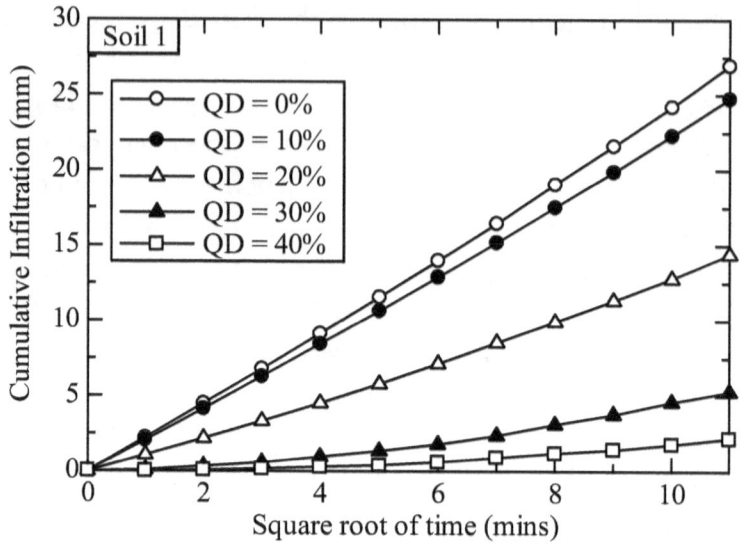

(b)

Fig. 5.4.2 Effects of Quarry Dust on Sorptivity (a), and (b) Cumulative Infiltration of treated soil, A

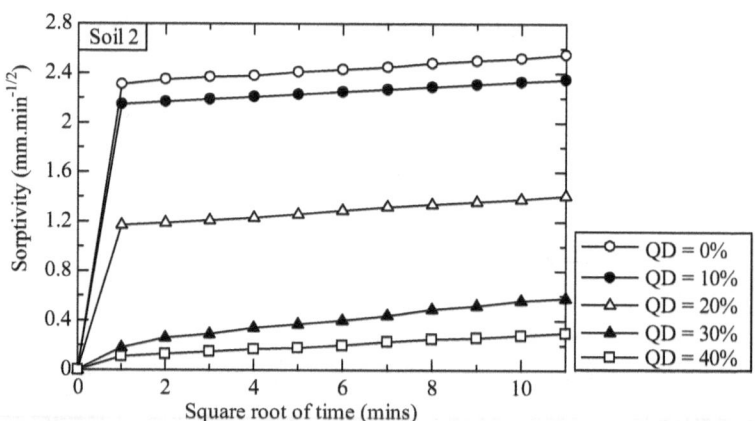

(a)

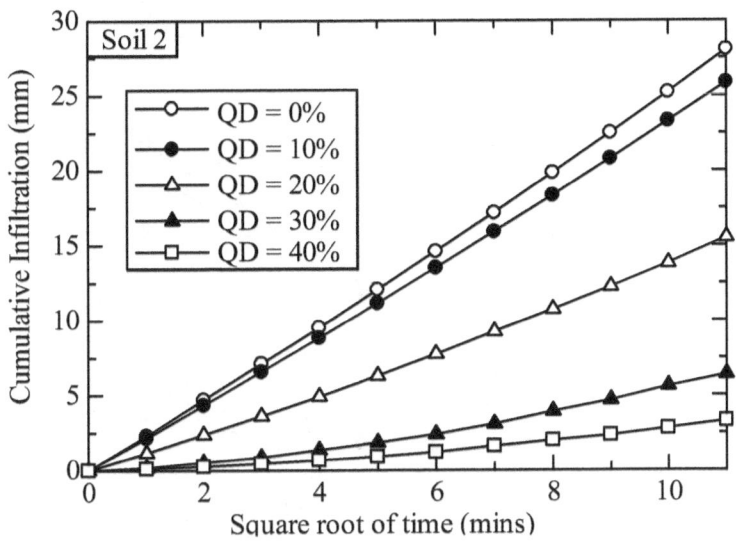

(b)

Fig. 5.4.3 Effects of Quarry Dust on Sorptivity (a), and (b) Cumulative Infiltration of treated soil, B

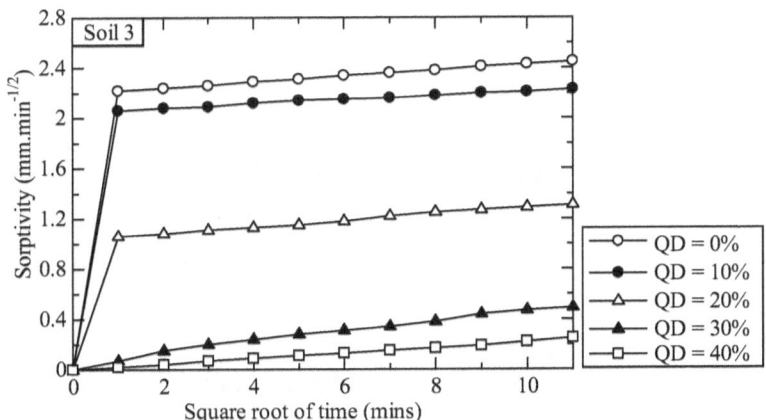

(a)

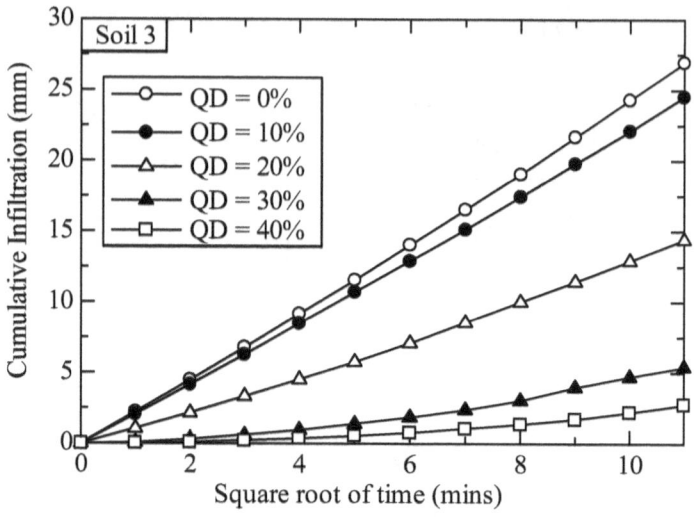

(b)

Fig. 5.4.4 Effects of Quarry Dust on Sorptivity (a), and (b) Cumulative Infiltration of treated soil, C

5.5 Erodibility

This is the technical measure of the erosive index of detachable soils exposed to erosion or surface runoffs. The average values of particle dislodgement (erodibility), which varied between 12.1g/min of the natural untreated soil and 5.2, 2.5 and 2.7 g/min for soils A, B and C respectively at 150% by weight addition of quarry dust based geopolymer cement (QDbGPC). The geopolymer cement additive and compacting energy may have reduced the erodibility potential. The reduction in the erodibility index for test soil A recorded an average value of 5%, 10% for test soil B and 6% for test soil C. These recorded rates of reduction were observed between 10% and 90 % by weight addition of QDbGPC. On higher addition of QDbGPC between 100 and 150% by weight, the average erodibility reduction dropped to 4%. So, the use of high geopolymer cement content should be encouraged because it enhances polymerization and calcinations reactions. However, the rate at which higher addition reduces erosion of soils is decreased. Geopolymer cement is highly resistant to shrinkage and eventually inhibits the formation of cracks. This in turn does not promote the ingress of moisture into the pavement structure or underlain subgrade because there are no voids created by lack of cracks. Consequently erosion process is reduced. The reduction in the average erodibility of the GPC treated soil may be due to the fine and reactive surface of the test materials. Also their ability to fill into the voids between soil particles to form floccs and densified matrixes goes a long way in reducing the action of erosive forces. The reduction in the average erodibility by the addition of the quarry dust base GPC showed the ability to improve durability of the treated soils under field condition. Field conditions are exposed to moisture effects or are hydraulically bound. Furthermore, the energy of compaction is another factor to be considered because erodibility depends partly on it. These results have been achieved without the application of ordinary Portland cement or any other conventional type of binders. Finally, a regression model on the erodibility potential of the quarry dust based geopolymer cement treated soils was conducted as presented in Fig. 5.5.1. The results showed that test soil A has a regression coefficient of 0.98, with test soils B and C having 0.94 and 0.96 respectively. This shows that test soil A has the best behavior with respect to the treatment protocol conducted with quarry dust based geopolymer compared to soils B and C. The overall erodibility potential behavior of the soils when treated with quarry dust geopolymer cement can be represented as follows in Equations 5.5.1, 5.5.2, and 5.5.3 for treated, improved and re-engineered soils A, B, and C;

$$E_A = -0.0401p + 12.029 \qquad (5.5.1)$$

$$E_B = -0.0669p + 11.171 \qquad (5.5.2)$$

$$E_C = -0.0683p + 11.727 \qquad (5.5.3)$$

Where,

E_A, E_B, and E_C = the erodibility potential for soils A, B and C and p = QDbGPC percentage by weight.

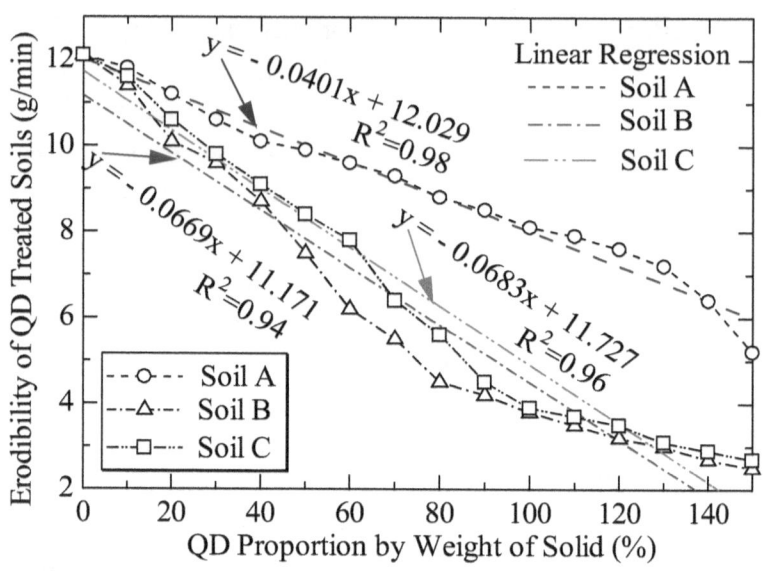

Fig. 5.5.1 Effect of Quarry Dust Proportion on Erodibility Potential of Treated Soil

5.6 California Bearing Ratio

CBR is a strength characteristic feature of soils used to determine its suitability to withstand shear deformations by punching and penetration. It establishes the overall thickness of pavements and other horizontal infrastructures. Because this has a direct relationship with the density of compacted soils, MDD is directly related to CBR. Natural and treated compacted soils are subjected to axial load and the axial strains are monitored. There are two conditions under which this is experimented. These are the soaked and unsoaked bearing ratios determined under 2.5 mm and 5.00 mm penetrations. As discussed earlier on compaction behavior of the treated soft soils, the ash materials and its derivatives improve on the density properties giving compactness that determines the California bearing ratio under axial loads. These additives improve on the soils ability to form a densified mass and compaction completes the process of densification and strengthening.

The first case is the California Bearing Ratio behavior of the crushed waste ceramics (CWC) treated soils A, B, C and D results are presented in Fig. 5.6.1. There was a steady strength improvement on the CBR of the treated soils with increase in CWC proportions with a maximum of 48%, 48%, 49% and 45% recorded on test soils A, B, C and D respectively at 120% addition of CWC by weight. These values which were greater than 20%, satisfy the materials requirement for utilization as stabilized sub-grade material on Nigeria roads. The consistent increase in the CBR with the addition of CWC could be due to the presence of sufficient release of calcium required for the formation of hydrated Calcium Silicates (CS) and Calcium Aluminate (CA). These are the major compounds responsible for strengthening. The soil + CWC mix met the basic California Bearing Ratio value of 20 – 30% specified by Onyelowe and Okafor for pavement materials suitable to be utilized as base course materials determined at Maximum Dry Density and Optimum Moisture Content.

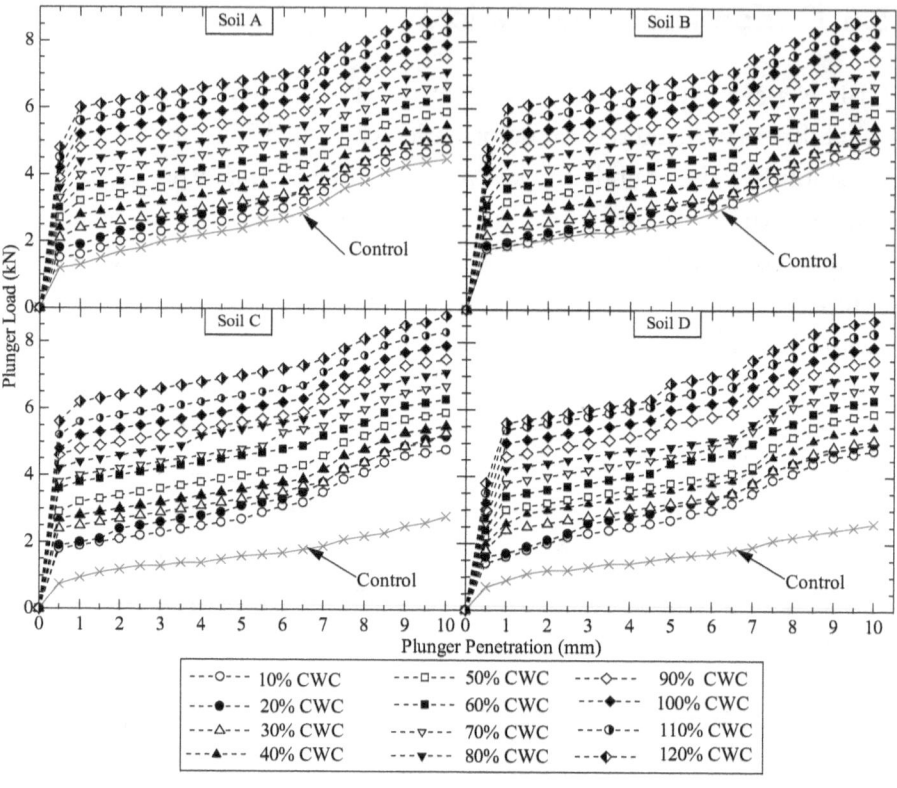

Fig. 5.6.1 Effects of CWC on California Bearing Ratio Behavior of Treated Soils

The second case in the stability experimentation is the observations of the effect of the crushed waste plastics (CWP) on the California bearing ratio (CBR) of the treated soils A, B, C and D are presented in Fig. 5.6.2. CBR is the measure of the ability of materials to resistance failure by shearing. It is an axial failure resistance measurement. The test soils had initial CBR values of 12, 13, 8 and 7% respectively achieved at maximum dry density and optimum moisture content. The soils were treated with CWP at the proportions of 10 to 120% in increments of 10% by weight of the dry solid. This admixture possessed a high content of aluminosilicates and exhibited pozzolanic properties with elastic properties. The blends between the test soils and CWP were compacted and tested for axial failure resistance. Test soils A, B and D showed similar behavior with the addition of CWP while soil C behaved somewhat differently. This is due to the fact that test soil C recorded the highest liquid limit, plastic limit and lowest shrinkage limit at the natural but disturbed state. The CBR of the treated test soils consistently improved with increased CWP content and recorded CBR values of 30, 38, 56 and 33% respectively at 120% by weight proportion of CWP. This shows a further potential of improvement on addition of CWP to the treated soil matrix. This behavior was as a result of the pozzolanic reaction between the aluminosilicate compounds of the additive and the test soils where the mixed matrix formed sequestrum and floccs with the additive. This reaction improved the calcination reaction and enhanced strength gain and densification. Cation exchange reaction within the double diffused layer of the mixed matrix of treated soils and additive also contributed to the strength improvement and resistance to shear failure of the treated soils.

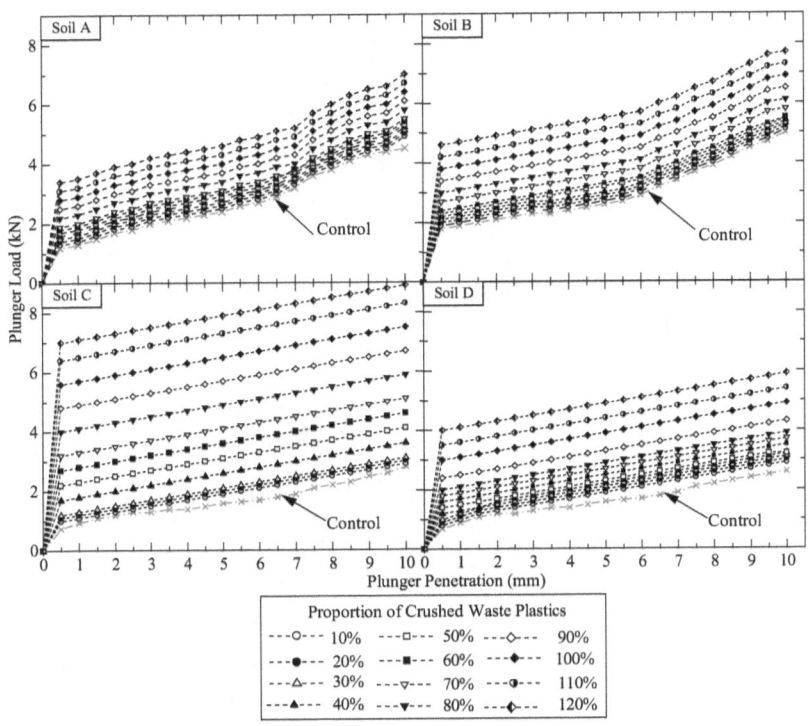

Fig. 5.6.2 Effects of CWP on California Bearing Ratio of Treated Soils

The third situation is the California bearing ratio behaviour of ordinary Portland cement (OPC) plus coupled composite of quarry dust based geopolymer cement (QDbGPC) treated soil with crushed waste glasses (CWG). CBR test was conducted to determine the untreated and treated soils resistance to shear failure when subjected to axial loads. Traffic loads are axial dynamic loads pavement facilities are exposed to from vehicles of various sizes and penetration pressures. Pavements and pavement foundation fail by shear or lateral displacement (deformation). Hence it is important to observe the rigidity or stiffness of subgrade materials used as underlain structures. The CBR behaviour results of the underlain subgrade soil were presented in Figs. 5.6.3 & 5.6.4 and Table 5.6.1. The studied soil was observed to be an expansive soil and was treated alternately with QDbGPC and OPC in the ratios of 0:0, 0:40, 5:35, 10:30, 15:25, 20:20, 25:15, 30:10, 35:5, and 40:0% by weight respectively. The effect of 4%, 8%, 12%, 16% and 20% by weight of crushed waste glasses over the cemented test soil was also observed. While 0:0% of the cements proportion by weight of solid served as the control point, the proportions of GPC increased from 5% in an increment of 5% while OPC decreased from 40% at the rate of 5% also. There was a consistent improvement on the CBR value of the treated soil with increased QDGbGPC and reduced DOPC proportions. These improved CBR values were greater than 20%, and satisfy the material condition for use as improved subgrade material on Nigeria's dilapidated roads. The consistently increased CBR values with the addition of QDGbGPC was due to the presence of adequate amount of calcium required for the formation of Calcium Silicate Hydrate (CSH) and Calcium Aluminate Hydrate (CAH), which are the major compounds responsible for the formation of sequestrum, floccs and strength development. The soil + QDbGPC blends at 40:0% by weight cementation met the minimum requirement for CBR value of 20 – 30% specified by Dogbey and Gidigasu for materials suitability for use as base course materials when determined at MDD and OMC. Increase in CBR value, was an indication of the improvement observed in MDD, which is attributed to the compatibility of the grains of soil due to the increased cations released and the high pozzolanic properties of the QDbGPC such that greater polycondensation and densification were achieved.

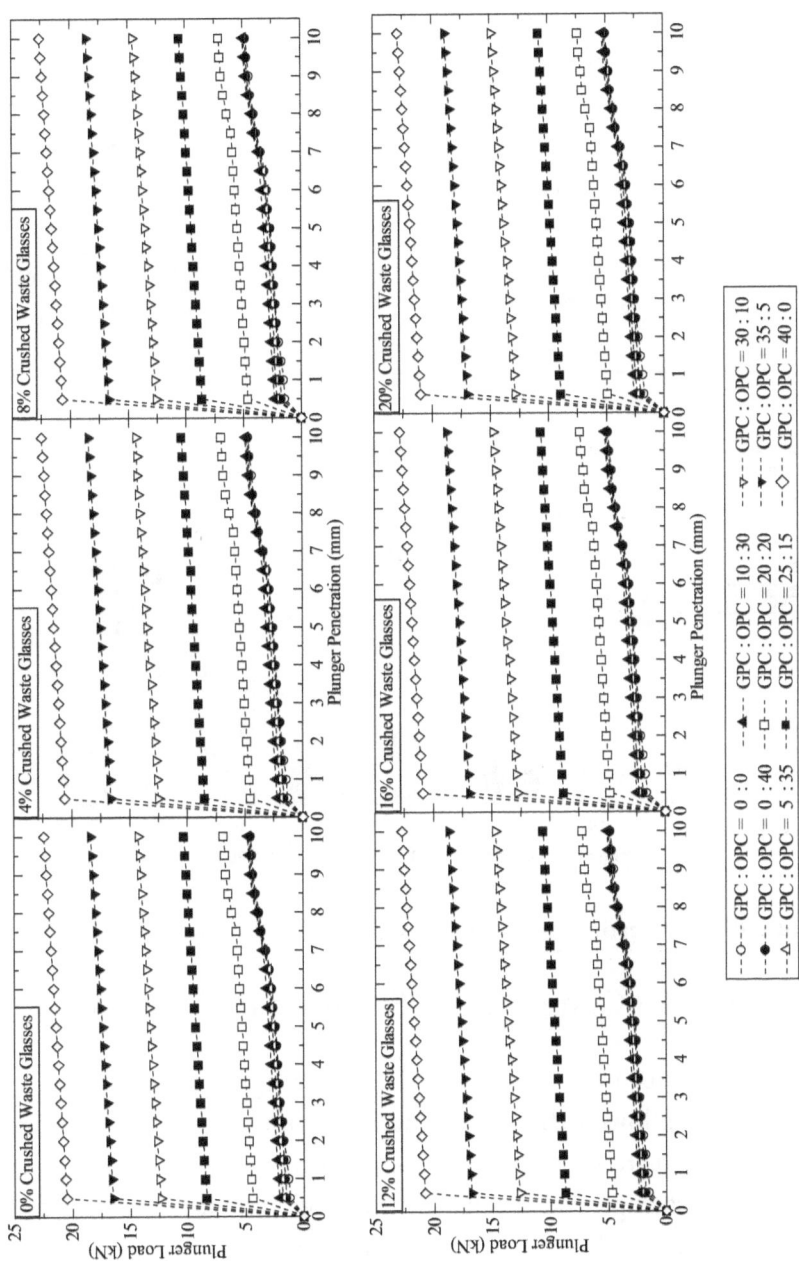

Fig. 5.6.3 Effect of Crushed Waste Glasses Proportion on the CBR behaviour of DOPC+QDbGPC (%) treated soil.

Table 5.6.1 California bearing ratio of OPC+QDbGPC (%) treated soil with CWG

CWG Proportion by wt (%)	CBR of OPC+QDbGPC (%) treated soil with CWG								
	0	0+40	5+35	10+30	20+20	25+15	30+10	35+5	40+0
0	13	14	17	19	36	66	97	128	158
4	14	15	17	20	37	67	97	128	159
8	16	16	18	20	38	68	98	129	159
12	17	17	20	21	39	69	99	130	160
16	17	18	20	22	39	69	100	131	161
20	18	19	21	23	40	70	100	131	162

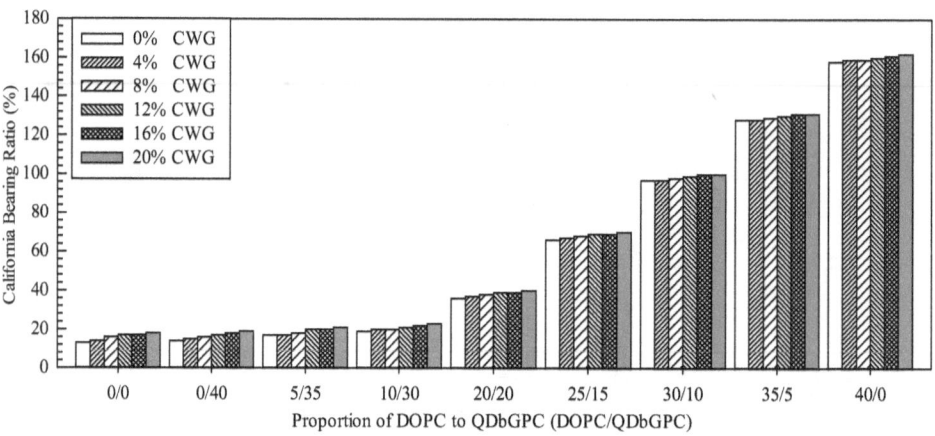

Fig. 5.6.4 California Bearing Ratio of OPC+QDbGPC (%) treated soil with CWG

5.7 Unconfined Compressive Strength and Durability

Compressive strength of soft soils is one the strength characteristics of that has equally allowed the determination of long-term performance and behavior of treated soils more especially under hydraulic influence through loss of strength on immersion method. The failure point of a treated and compacted sample under compression is determined with respect to the sample surface area. The formation of calcium silicate hydrates and calcium aluminates hydrates with addition of ash, powder and ash/powder based geopolymer cements in soft soils promotes strengthening of treated soils. This is due to those hydrates responsible for strengthening and this improves the compressive strength of the treated samples. The innovation here is that these hydrates of strength are formed utilizing ecofriendly materials derived from solid waste hence with zero release of carbon oxides to the environment. The durability potential is the compression index between completely open air cured samples and partly open air cured and partly full immersion cured specimens. As has been captured previously, these amorphous materials, powder and geopolymer cements have shown to resist moisture influence by reducing its effect on soft soils and improving on the volume changes that impair the overall long-term performance and service of the infrastructures constructed with these materials. Similarly, flexural strength is the magnitude of stress and force an unreinforced concrete structure can resist when subjected to bending or transverse rupture force which is also the material property known as the stress in a material just before failure or rupture. Previous experimentation in construction and mechanical materials sciences and engineering, in recent times, are involved in utilizing agricultural, household, municipal or industrial solid wastes to either partially or completely substitute conventional materials of concrete to achieve environmental friendly engineering goals. The blending of agricultural by-products as supplementary cementitious additives has been studied with positive results in the production and application of blended concrete and other coupled composites. This results not only product flexure resistant structural elements but also durable in terms moisture, sulfate, heat, and crack long-term resistance and sustainable in terms of available and workability of these additive materials. The improvement recorded in this property on treated soils and concrete and asphalt has been consistent and invaluable. These materials either serve as modifiers by improving on the microstructural properties of the elements or as fillers by improving on the porosity of the compacted structural elements.

The first situation is the compressive strength behavior of the coupled composite quarry dust based geopolymer cement treated lateritic soils cured for 28 days was presented in Fig. 5.7.1. The graphs show an almost similar behavior of the soils under the QDbGPC treatment. It was important to cure for the long period of 28 days to ascertain the behavior of the treated soils under long period of exposure to moisture. The treated soils A, B and C blended with the geopolymer cement remarkably improved in the compressive strength. The high proportion of aluminosilicates contained in the blended and coupled materials to synthesize the geopolymer cement may have contributed to this behavior. This possessed the ability to reduce adsorbed moisture at the polymerization process thereby causing CH soils to perform like granular soils. This also may be attributed to the physicochemical and high pozzolanic characteristics of the geopolymer cements. The compounds of the synthesized geoplymer cement, at high concentration may have equally increased the contact force or interconnection between soils particles to produce a highly homogenous material.

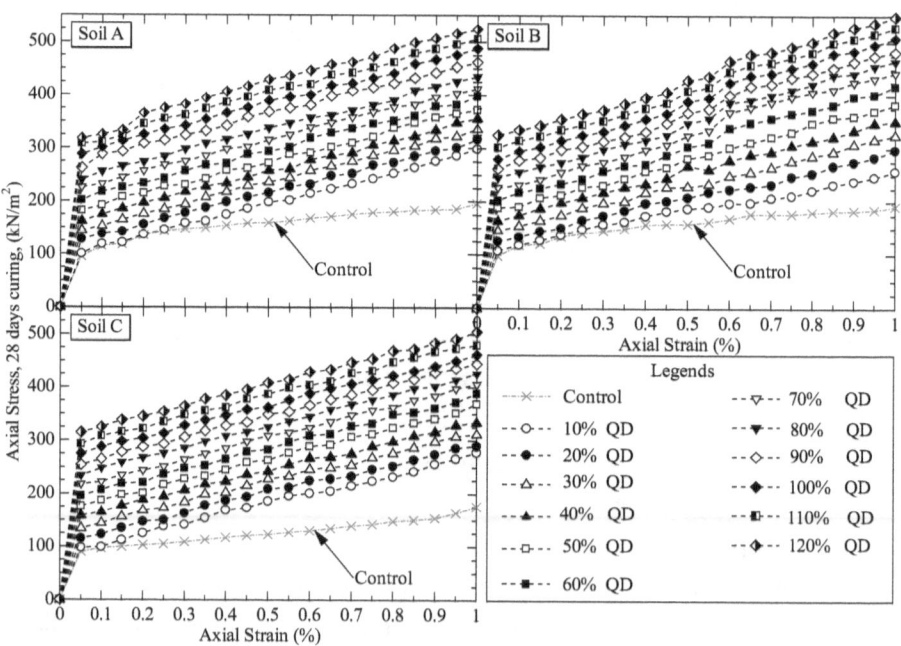

Fig. 5.7.1 Unconfined Compressive Strength Behaviour of Treated Test Soils at 28 days curing: Soil A, Soil B, and Soil C

The loss of strength on immersion test results and the associated durability index of quarry dust based geopolymer cement (QDbGPC) treated lateritic soils are presented in Fig. 5.7.2. It presents the behavior of three different test lateritic soils treated under laboratory conditions. The initial test on the untreated test soils, which served as the reference test showed that the compacted natural soil was not durable. It showed a durability index of 76.10%, which was less than the standard minimum durability requirement of 80%. On the addition of varying rates of quarry dust base geopolymer, the durability potential improved considerably and exponentially. This was with reference to the open-cured and completely immersion cured sets of treated and untreated specimens. The matric suction on immersion affected the strength development of the treated soils, but the results remained within the minimum durability index requirements. This could have been due to the fact that the geopolymer materials showed high pozzolanic characteristics. This resulted from the polymerization of the coupled elements of the synthesized geoplymer cement, which are highly resistant to moisture attack. Also, this behavior was because of the cations released at the chemical reaction interface that generated a strong resistance to the effect of moisture ingress on immersion. Moreover, the reaction between the soil anions and the geopolymer materials cations at the chemical reaction interface contributed to the formation of fluccs and progressive densification and gain in strength. This behaviour also resisted the effect of moisture intake at the immersion of the test specimens. The building up of the pozzolanic reactions by greater polymerization due to geopolymer cement increase improved strength development in the test specimens at the different rates. Hence, the property led to an increase in durability potential.

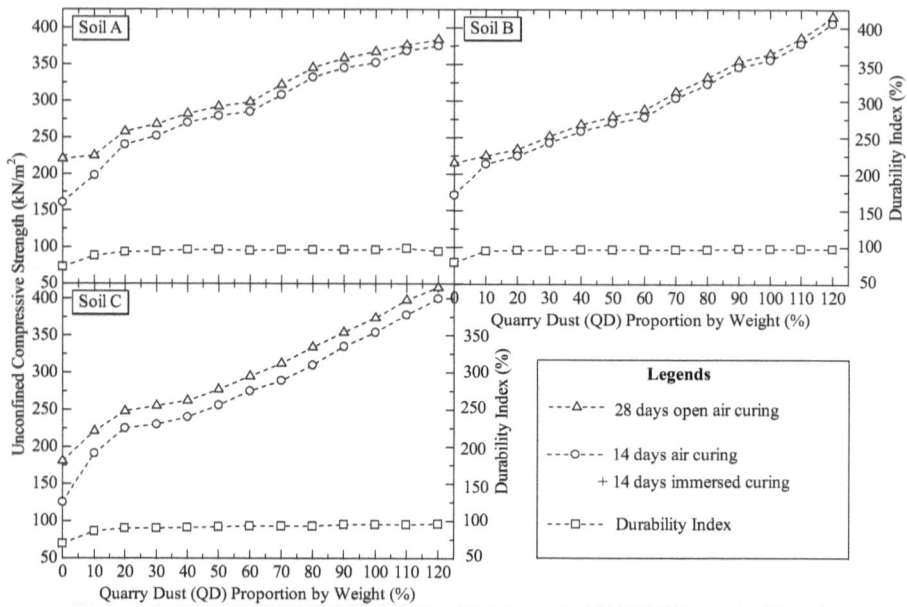

Fig. 5.7.2 Effect of Quarry Dust Propotion on Compressive Strength Loss on Immersed and Durability Index of Treated Test Soils: Soil A, Soil B, and Soil C

The second situation is the test soils A, B and C that were treated with varying proportions of clear quarry dust (QD) in the rate of 10%, 20%, 30%, 40%, 50%, 60%, 70%, 80%, 90%, 100%, 110% and 120% by weight. Test specimens were prepared and subjected to compression tests. Fig. 5.7.3 presents the deformation behaviour of the treated samples under the addition of quarry dust (QD) and cured for 28 days when subjected to axial loads. It is observed that soils A and C have a somewhat similar deformation behaviour under the same axial loading. The tests showed a consistent improvement in the compressive strength of the treated soils. The consistent strengthening was due to the formation of floccs at the double diffused layer of the treated soil with the dissociated ions from the aluminosilicates forming calcium aluminate hydrates responsible for strength gain in treated soft clayey soils. Secondly, the cation exchange reaction encouraged the formation nucleating surface for strength gain.

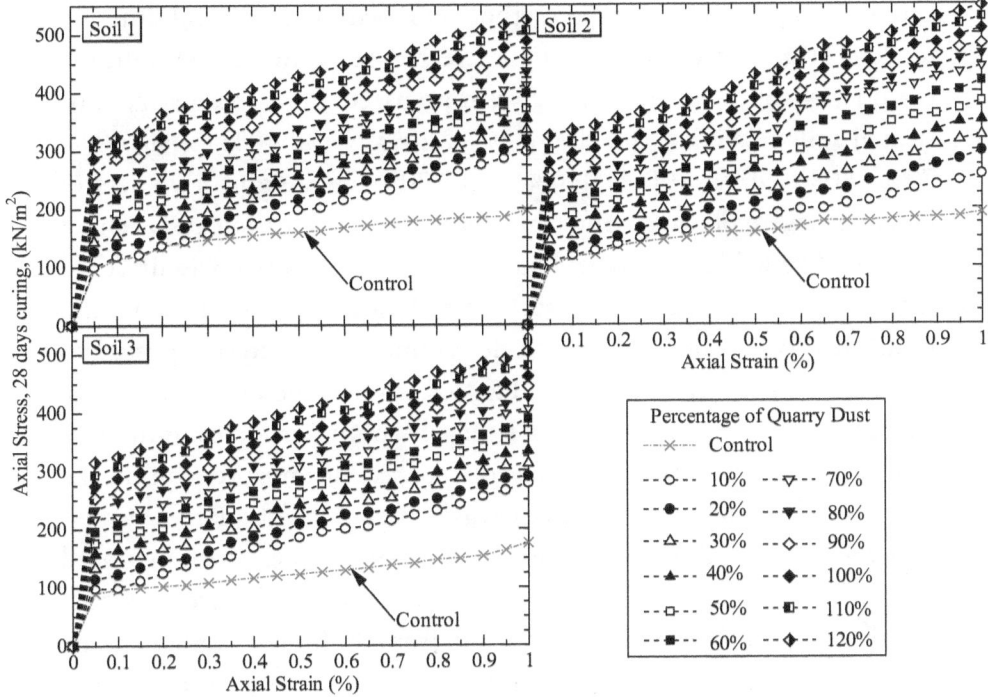

Fig. 5.7.3 Unconfined Compressive Strength Behaviour of Treated Test Soils at 28 days curing: A, B, and C

Loss of strength on immersion method was used for the second situation to determine the durability of the treated soils under hydraulically bound environment. Specimens were prepared from the quarry dust treated soils treated at the rate of 10%, 20%, 30%, 40%, 50%, 60%, 70%, 80%, 90%, 100%, 110% and 120% by weight. Two sets of specimens were prepared from this treatment specification and each set replicated to achieve average results. The first set was open-air cured for 28 days while the second set was open-air cured for 14 days and immersed for another 14 days to determine the effect of exposure to moisture on the treated samples. Fig. 5.7.4 shows the behaviour of the treated samples exposed to loss of strength on immersion examination. The compression test results showed a consistent improvement on the strength of the treated soil with increased variations of quarry dust. So also, the recorded durability index showed improvement with increased quarry dust addition. This is due to the formation of the hydrates with the ions of aluminium, silica and calcium from the highly aluminosilicate quarry dust ion dissociation and also the cation exchange reaction. The formation of floccs also contributed to the consistent strengthening of the treated soils A, B and C. There are clear indications on the behaviour of the fully 28 day air cured treated specimens and the 14 day air cured and 14 day immersion cured treated specimens. Beyond 100% by weight addition of QD, the treated test soils, the effect continued to improve on the compressive strength though the difference between air cured and partly immersion cured was maintained throughout the experimentation. The consistent improvement in the strength property was due to release of more aluminosilicates from the QD upon the addition increasing and continuing the carbonation, calcination, pozzolanic and hydration reaction and also the cation exchange reactions binding the fine particles of the test soils. This binding process is called flocculation which gives rise to the formation of C-A-H, and C-S-H from the QD oxides compositions upon the addition of moisture. Also, prolong exposure to moisture also prolong the hydration reaction with the cementitious additive yielding strengthening of the treated soils. This characteristics was also responsible for the improved durability indexes.

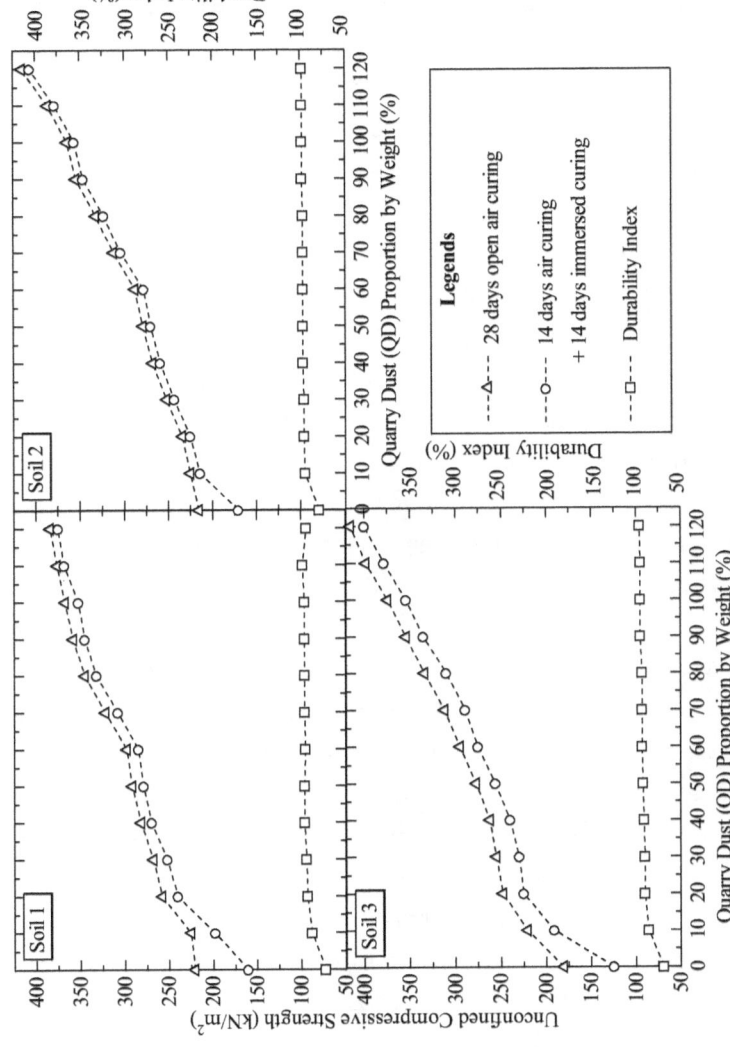

Fig. 5.7.4 Effect of Quarry Dust Proportion on Compressive Strength Loss of Specimens Immersed and Durability Index of Treated Test Soils: A, B, and

5.8 Resistance Value

The resilient modulus and resistance value are engineering properties used to characterize materials in unbound construction environments especially in pavement construction. This measures the engineering materials stiffness (subgrade stiffness) and provides a method to analyze and study the stiffness of materials; treated and untreated under different loading and environmental conditions. Moisture, density, stress, temperature, vertical and lateral loading, etc. are some of those conditions. Most times, foundation materials subjected to vertical loading suffer lateral deformation and the ability to withstand this form of failure is known as resistance value (R-value). It is important to also note that these are strength properties and depend on density attained on an optimum moisture content. The materials additives derived from solid waste (ash, crushed waste (powder), and coupled geopolymer cements or other composite materials) react with the minerals of soft clay soils in a treated blend to form floccs and sequestrates. Experimental results have shown that these materials improve the resistance value, deviatoric stress and resilient modulus of the treated soft soils to achieve the best performance as eco-efficient and sustainable infrastructures. Similarly, the Marshall stability is the measure of the optimum binder content in bituminous pavements however in this case modified with derivatives of solid waste. Marshall Asphalt concrete mixtures are designed and produced to achieve standard specifications in terms of stability, flow, porosity, and density characteristics. The relative impermeability of asphalt mixture depends on the quality or content of the binding materials. The asphalt concrete that meets the marshal stability requirements is always desirable and this depends on the quality of the binder. The conditions of an optimum binder content that balances different design mixtures should be taken into consideration and again the emissions released into the atmosphere is of utmost concern in this work. The incorporation of geopolymer cement synthesized from solid waste ash materials into the production of asphalt as modifiers reduces the amount of oxides of carbon emission and produces a homogenous blend with high resistance to sulfate attack, and crack. Secondly, these materials as ash or powder have been used also as fillers to improve on the porosity and flow of the concrete. Experimental results have shown tremendous results achieved in this line.

The first case is the results of the stabilometer test adapted with the modified triaxial compression test set up for the CWC treated test soils are presented in Figs. 5.8.1, 5.9.1, 5.10.1 and 5.11.1. This was conducted on the treated soils to determine the treated soils ability to resist lateral deformation when subjected to both horizontal and vertical pressure loading of the stabilometer under laboratory conditions. The results showed a consistent reduction in the deformation over a consistent improvement on the resistance value of the treated soils subjected to the dual loading arrangement. The deformation on the treated soils increased with increased pressures but reduced with increased addition of the CWC. The R-value also improvement consistently with increased proportions of the additive but reduced under increased pressure (vertical and horizontal) of the stabilometer. The index of reduction of the deformation with increased CWC was recorded at 23% for test soil A, 21% for test soil B, 24% for test soil C and 28% for test soil D. similarly, the improvement index on the r-value of the treated soils were recorded as 34%. 32%, 36% and 38% respectively with increased proportions of CWC by weight. This is behavior was due to the ability of the pozzolanic compounds and the aluminosilicate strength of the additive to for floccs and densified matrix compacted at optimum moisture content. The crystalline particles of the CWC filled the pore spaces of the treated soils thereby improving the porosity of the soils matrixes hence achieving high resistance value test soil materials suitable as subgrade and sub-base pavement materials of high stiffness. The measure of the failure of pavements is based on both shear and lateral failure. While CBR and resilient modulus gives a clue to monitoring the failure of pavements by shear, the r-value gives an expert guideline of the resistance of the pavement material to displacement.

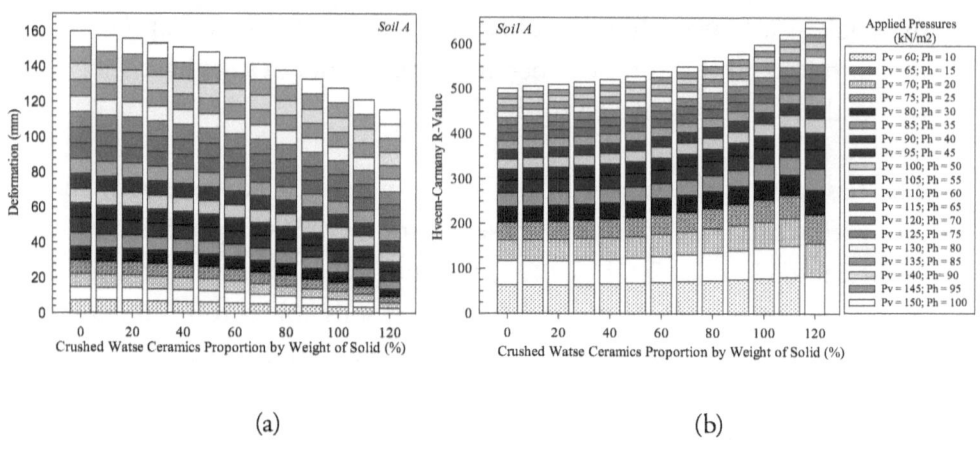

Fig. 5.8.1 Effect of CWC on Deformation (a) and R-Value behavior of Treated Soil A (b)

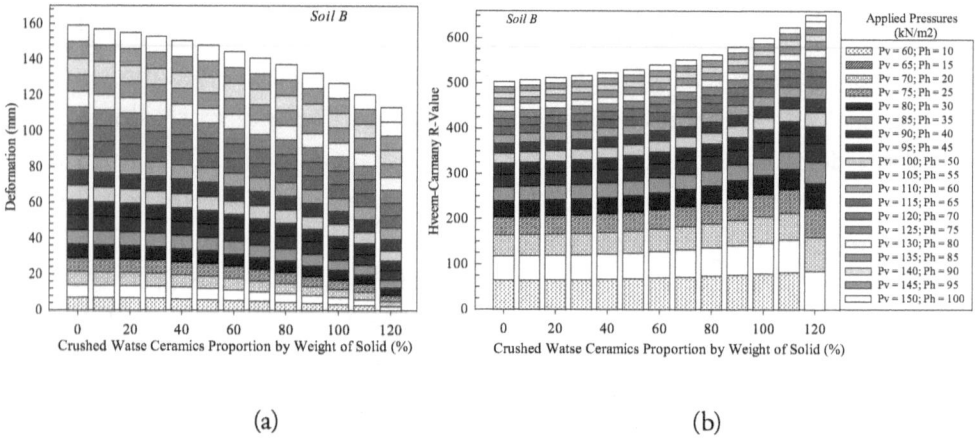

Fig. 5.9.1 Effect of CWC on Deformation and R-Value behavior of Treated Soil B

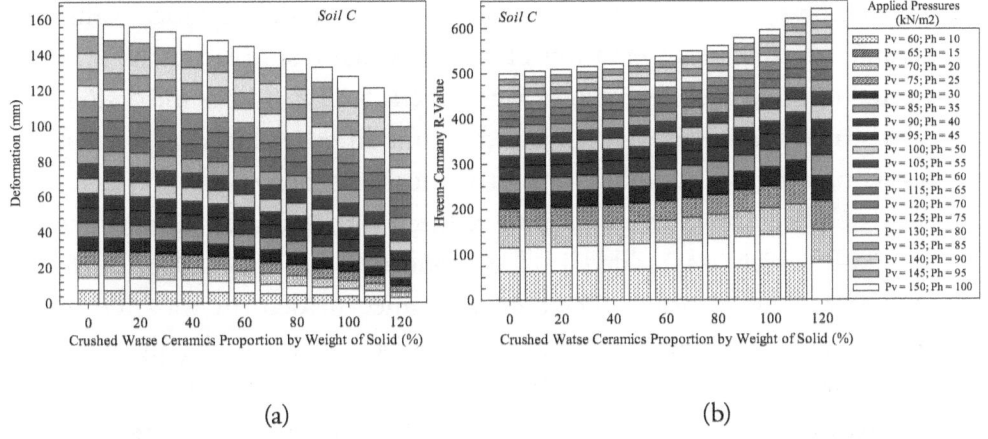

Fig. 5.10.1 Effect of CWC on Deformation and R-Value behaviour of Treated Soil C

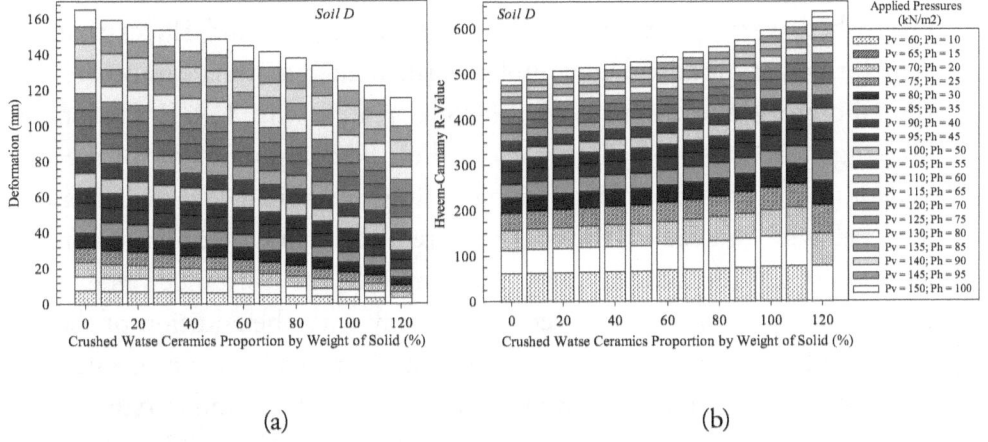

Fig. 5.11.1 Effect of CWC on Deformation (a) and (b) R - Value behavior of Treated Soil D

The second case is the lateral deformation and resistance value laboratory exercise observation of the crushed waste plastics (CWP) treated test soils are presented in Figs. 5.12.1, 5.12.2, 5.12.3 and 5.12.4. The treated specimens were subjected to a modified triaxial compression test consisting of a combination of vertical and horizontal pressures. The lateral deformation consistently reduced with increased proportions of CWP which was added in the proportion of 10%, 20%, 30%, to 120%. The recorded increased resistance to deform was due to the formation of silicates and aluminates of calcium under hydrated laboratory conditions. This also led to the formation of densified matrixes of treated specimens that defied the effect of the pressures that they were subjected to. This goes to show that the treated soils achieved the ability to resist lateral cyclic loads like traffic loads when subjected to the highway traffic conditions with the increased addition of crushed waste plastic materials. The sequestrum and floccs that were formed at the adsorbed complex of the treated materials improved the laboratory performance of the soils treated with CWP. Cation exchange reaction also caused a released of the compounds of silicates and aluminates responsible for strengthening in stabilized treated soils. This behavior also caused the consistent reduced lateral deformation on addition of the admixtures by weight. Similarly, the resistance value of the four test soils were improved consistently with the addition of the CWP under the applied pressures of the modified triaxial compression condition. The hydration reaction at optimum moisture and maximum dry density of the treated matrixes was responsible for the buildup of the resistance value of the treated soils. Also, polymerization reaction of the plastic compounds with the test soils when mixed with the optimum moisture condition help the soils achieve a consistent improvement in the resistance value (r-value). Resistance value is dependent on the lateral deformation. The foregoing implies that resistance value is inversely proportional to the lateral deformation, which further implies that when lateral deformation reduces as a result of the addition of geomaterials responsible for strength gain, the resistance value improves. This is the case in the present work. It is novel to achieve an improved r-value with a geomaterial from solid waste.

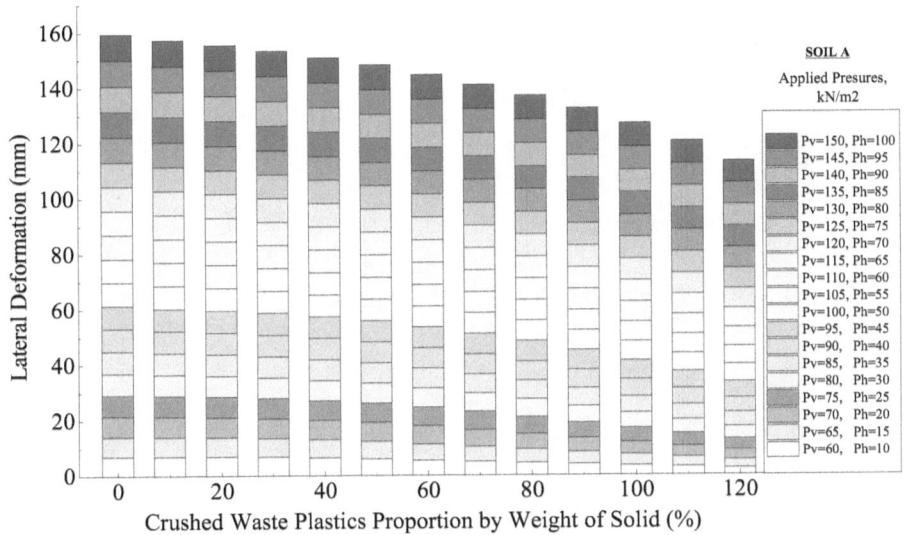

(a)

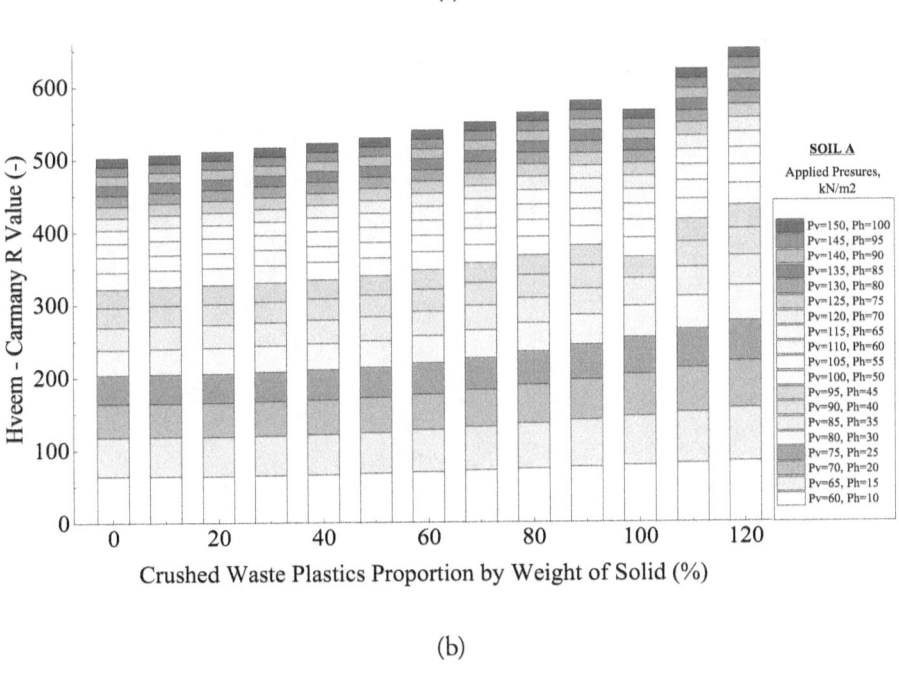

(b)

Fig. 5.12.1 Effect of CWP on lateral Deformation (a), and (b) R-value behavior of Treated Soil A

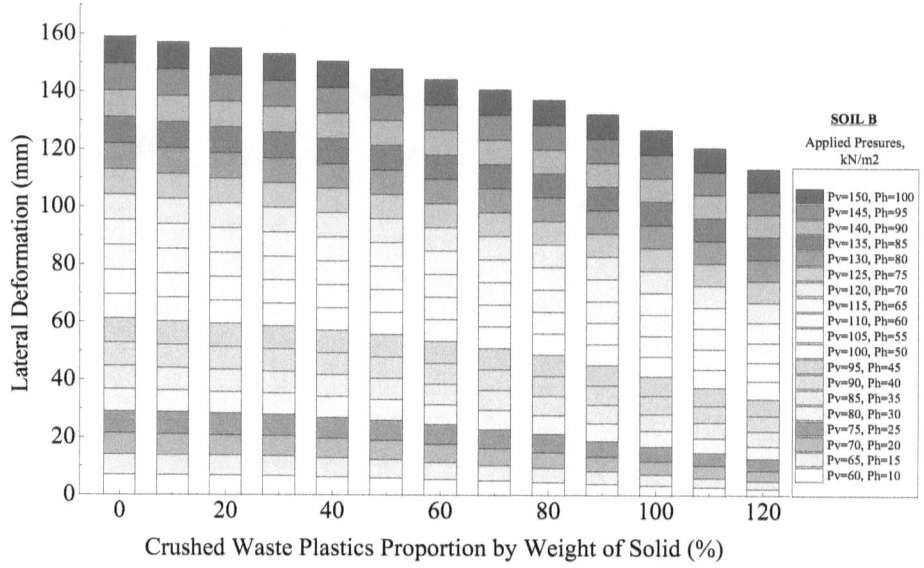

(a)

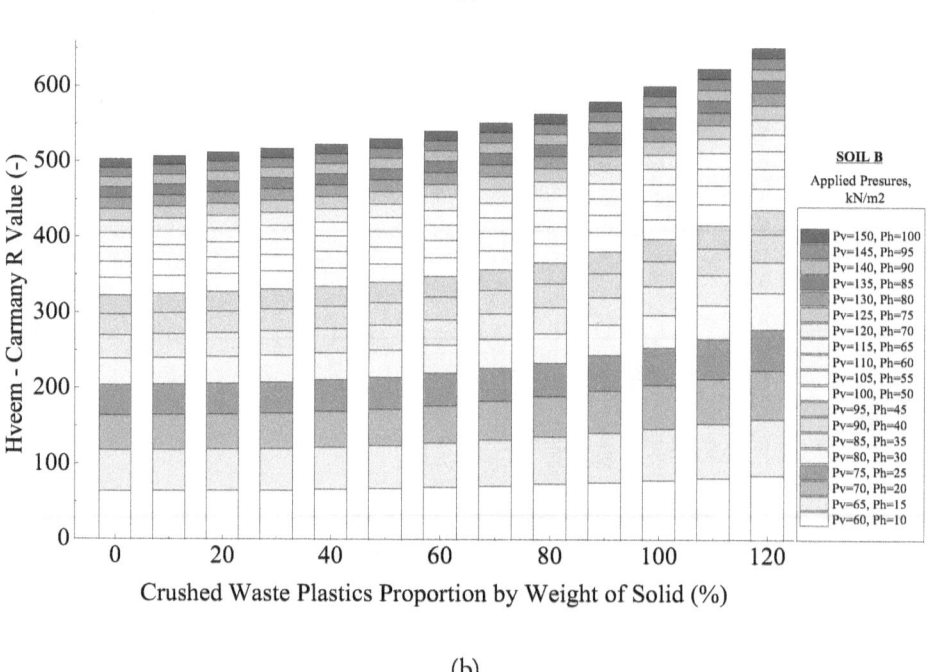

(b)

Fig. 5.12.2 Effect of CWP on lateral Deformation (a), and (b) R-value behavior of Treated Soil B

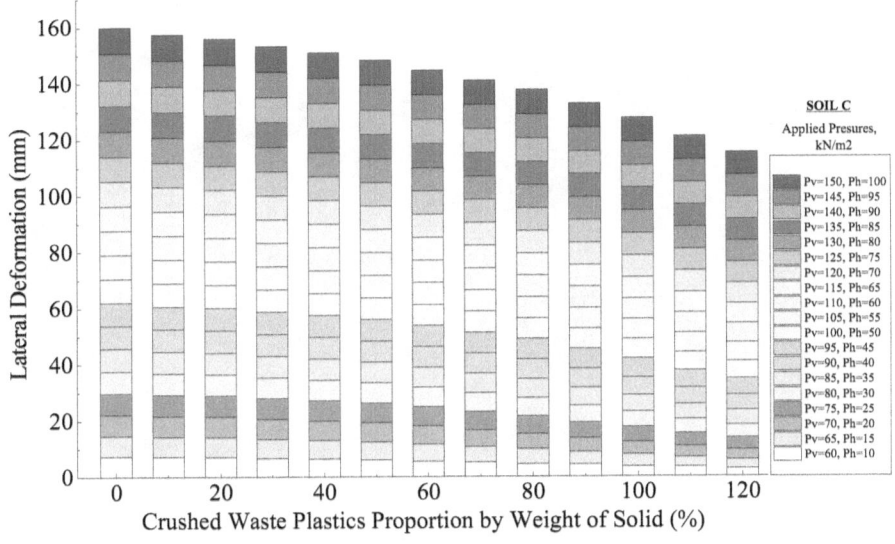

(a)

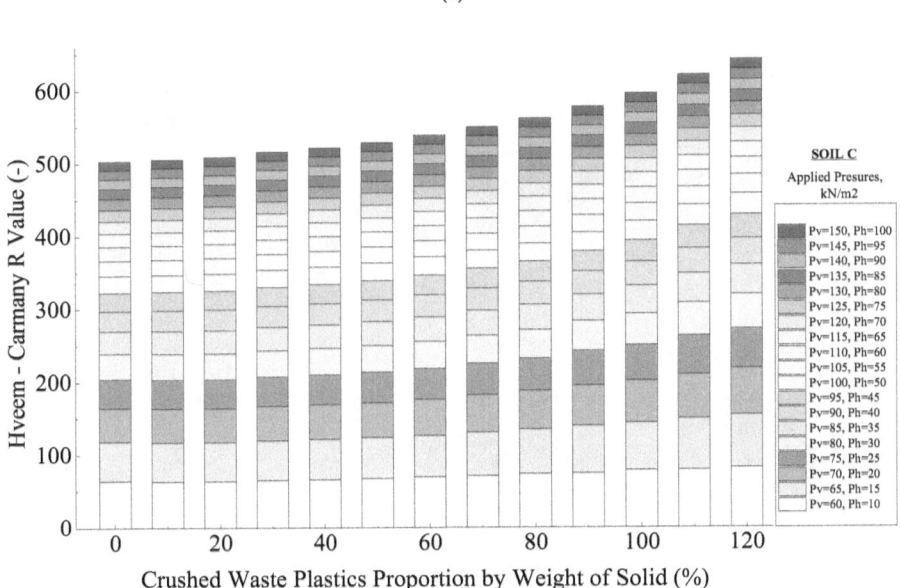

Fig. 5.12.3 Effect of CWP on lateral Deformation (a), and (b) R-value behavior of Treated Soil C

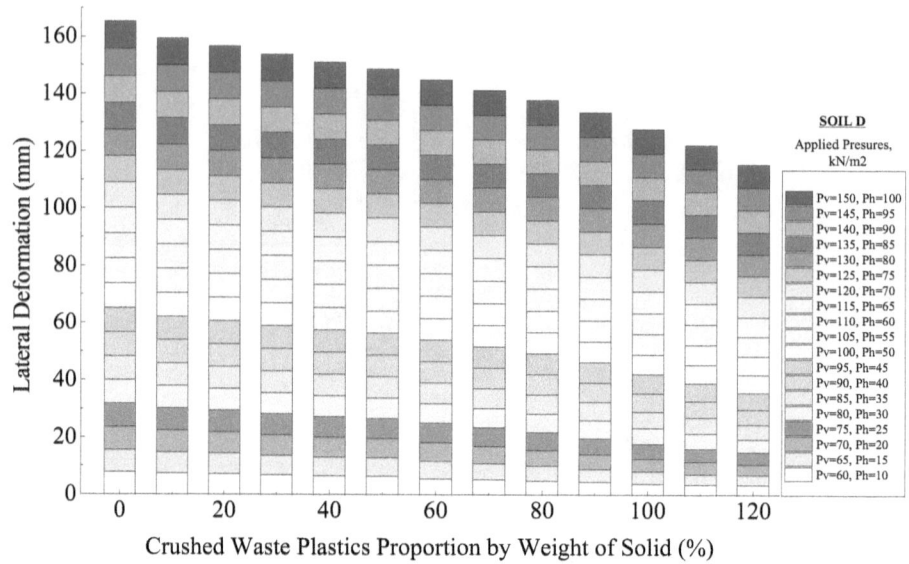

(a)

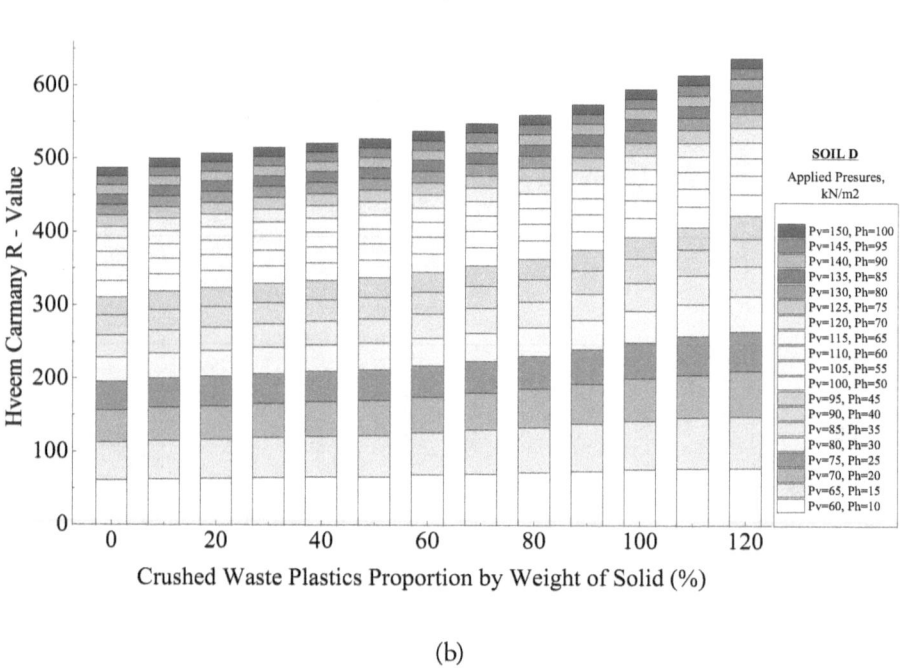

(b)

Fig. 5.12.4 Effect of CWP on lateral Deformation (a), and (b) R-value behavior of Treated Soil D

5.9 Resilient Modulus

The results of the resilient modulus and its corresponding deviatoric stress of the treated soils used to characterize the treated matrix as a subgrade materials is presented in Figs. 5.9a and 5.9b. The applied deviator stress and the recoverable strain of the modified triaxial test on the treated specimens were used. The four test soils behaved in almost the same pattern with similar reactions with increased crushed waste ceramics (CWC). The deviatoric stress consistently increased with increase in the proportion of the admixture for test soils A, B, C and D. It is important to note at this point that the additive CWC is a highly aluminosilicate compound with a crystal texture prior to its utilization in the stabilization procedure. These compounds are responsible for pozzolanic reaction, and strengthening by forming silicates of calcium hydrates and aluminates. Test soil A, B and C had an improvement index of about 21%, while test soil D had an improvement index of 25%. The higher improvement index recorded with test soil D is in line with its natural soil high resilient modulus of 0.72E+05 which was improved upon. The hydration reaction between compounds of strengthening from the additive and the dissociated soil ions in contact with moisture had caused the improvement on both deviatoric stress and resilient modulus of the test treated soils. These results were recorded under cyclic loading on specimens subjected to testing sequences. The physical conditions that affect the resilient modulus (moisture and unit weight) were influenced by the introduction of the highly aluminosilicate CWC hence improving the strength behavior of the treated soils.

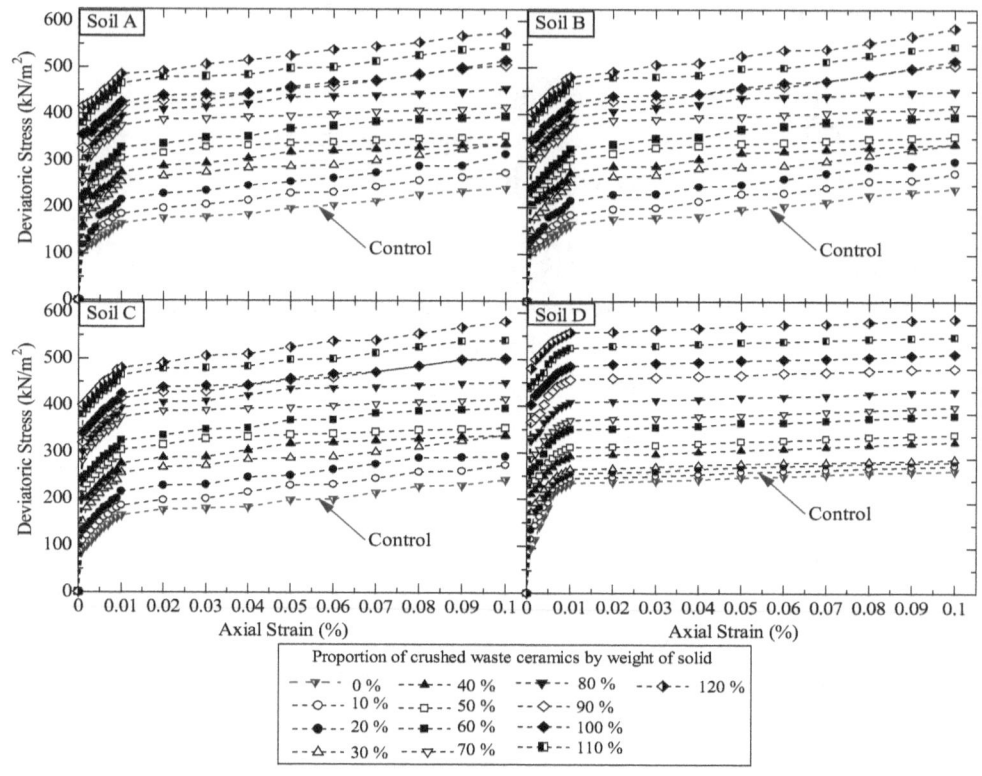

Fig. 5.9a Effects of CWC on Deviatoric Stress of the Treated Cemented Soils

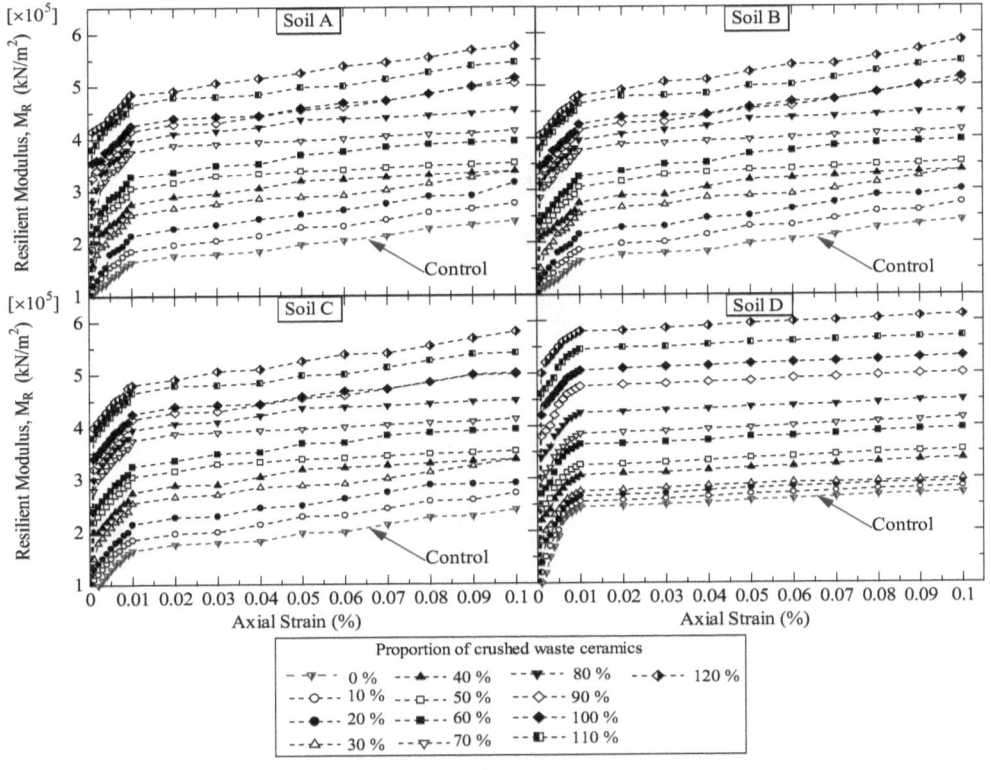

Fig. 5.9b Effects of CWC on Resilient Modulus, MR, of the Treated Cemented Soils

CHAPTER SIX

MODELING AND OPTIMIZATION TECHNIQUES IN SOILS RE-ENGINEERING

6.1 Analysis of Variance

The behaviour of the stabilized lateritic soils has been evaluated by different researchers under different loading conditions in the laboratory through various analytical and numerical methods. Right from the rise of soil stabilization and soil strength improvement in the field of Geotechnical engineering, researchers and experts have adopted many methods and approaches in the stabilization operation. From the era of purely mechanical stabilization to mechanical plus chemical to mechanical plus chemical plus biodegradable additive or the byeproduct of biodegradable additives; ash materials, soil stabilization has improved in various ways.

Lateritic soil as a construction material that plays a vital role in the field of Geotechnical engineering, highway engineering and civil engineering as a whole has been faced with diverse techniques and technologies aimed at improving its quality and property for the purpose of efficient engineering service delivery. As a result, many more methods have been adopted to certify the veracity of some of the practical, numerical, analytical, etc. methods. One of those employed in recent times is the analysis of variance. The lateritic soil stabilization adopts the method of treatment of the soil matrix with varying degrees of additive in the laboratory and the behavioural change in the Geotechnical properties of the treated soil observed to determine the best results suitable for the engineering operation of choice. Through the analysis of variance approach, different degrees of treatments carried out on the soil are evaluated for validity and possible interaction between its components to arrive at a certain degree of acceptance. It is through interaction between the components of the treated soil that the chemical reactions that take place like the cation exchange, hydration, flocculation, double diffused layer build up, etc. can possibly take place and leading to densification and hardening or strength gain of the stabilized matrix. So, the test for this interaction proves to be very vital to ascertain the conclusions being made at the end of a stabilization process and validate the decisions we make. Test of null hypothesis is a test that leads to a decision to accept or reject the hypothesis under consideration in an engineering design and its allied disciplines. Two hypotheses are involved, (i) *Ho*, which is the null hypothesis and (ii) *HA*, which is the alternative hypothesis. If *Ho* is false, then *HA* is true and vice versa. In carrying out a test, we may erroneously reject a hypothesis we ought to have accepted and vice versa. There are two types of errors that could be committed in this operation; (i) Error Type A: when one rejects *Ho* when it is true and the probability of committing this error is α and (ii) Error Type B: when one fails to reject *Ho* when *HA* is true and the probability of committing this error is $1-\alpha$, where α is the level of significance. The three methods of hypothesis testing in this soil stabilization research operation can be formulated. ANOVA Nonparametric rank transformation method by Kruskal-Wallis Test is the hypothesis testing method developed by Kruskal and Wallis in 1952 where the experimenter may wish to use an alternative procedure to the F test analysis of variance that does not depend on the assumptions of F test. K-W test is used to test the null hypothesis *Ho* that n treatments on a sample is identical against the alternative hypothesis *HA* that some of the treatments generate observations that vary from others. This is a nonparametric alternative to the usual ANOVA. To perform this operation, the observations $Y{ij}$ are ranked in ascending order and each observation is replaced by its rank, say $R{ij}$ and let $R{i}$ be the sum of ranks in the ith treatment.

6.2 Multiple Regression and Nonlinear Multiple Regression

This model studies the regression relationship on the stabilized lateritic soil-ash mixture between the Subgrade Stiffness (E) as the dependent variable and California bearing ratio (R), additive sample percentage by weight proportion (A) and moisture condition value (V). These are the independent variables in this prediction model. The strength and durability of hydraulically bound materials (HBM) for example materials used as sub-structural foundations (pavement subgrade) depend on the stiffness of the materials; treated or untreated. And the essence of treating these materials for the purposes of construction is to improve on the strength characteristics. However, developing models that predict and monitor the strength development and performance of the structures justifies the main concern of this work. This is important because in the developing countries of the world, pavement failures as well as structural foundation failures are common and the only procedure towards forestalling such occurrences is to develop a more robust mathematical relationship that incorporates the effect of the commonly used procedure of admixtures stabilization and moisture condition value; that takes care of the hydraulic exposure of the materials. The relationship between these variables is a nonlinear objective function. A nonlinear multiple regression relationship was formulated to express the relationship between the stabilization variables and the subgrade stiffness (E). The formulated nonlinear equation was linearized, solved and calibrated or correlated with the analytical expression of the subgrade stiffness according to Powell *et al*. And the correlation exercise showed that the predicted subgrade stiffness results have proven to possess a better correlation than the analytical expression, hence should be better to be applied in the design, construction and monitoring of the performance of transportation pavements. And as a general remark, the robust nature of the prediction will make it possible for the subgrade stiffness or any characteristic parameter of a treated hydraulically bound material like the case of substructures subjected to moisture effects to be evaluated by adjusting the equation to different degrees of admixture content, bearing ratio and moisture condition. Hence, this offers designers an opportunity to also monitor the behavior of the foundational subgrade and pavements performance. This mathematical procedure can be adapted engineering problems in Civil, Mechanical, Chemical, and Agricultural Engineering as well as allied disciplines.

6.3 Scheffe's Method for Sustainable Soils Re-Engineering

The method of optimization that will be used for the investigation is scheffe's simplex Lattice method for mixtures, where the property studied depends on the component ratios only. Firstly, a simplex is defined as a convex polyhedron with (k+1) vertices produced by k intersecting hyper planes in k-dimensional space. Any co-ordinate system above 3-dimensions are referred to as hyper planes, such planes are not orthogonal. A 2-dimensional regular simplex is, therefore, an equilateral triangle, while a 3-dimensional regular simplex is a regular tetrahedron. Scheffe used a regular (q − 1) − simplex to represent a factor space needed to describe a response surface for mixtures consisting of several components. If the number of components is denoted by q, then for binary system (q = 2) the required simplex is a straight line; for q = 3, the required simplex is an equilateral triangle; and for q = 4, the simplex is a regular tetrahedron. The response surface for such a multicomponent system is normally described by means of a high degree polynomial, of the type of Eq. 6.4.1, having number of coefficients given by n where n is the degree of the polynomial.

$$\hat{y} = b_0 + \sum_{1 \leq i \leq q} b_i x_i + \sum_{1 \leq i < j \leq q} b_{ij} x_i x_j + \sum_{1 \leq i < j < k \leq q} b_{ij} x_i x_j x_k + \sum b_{i_1 i_2 \ldots i_n} x_{i_1} x_{i_2} \ldots x_{i_n} \qquad (6.4.1)$$

6.4.1 Factor Space in Scheffe's Simplex Design

Simplex is the structural representational shape of a line or planes joining assumed positions of constituent materials (atoms) of a mixture. Scheffe considered experiments with mixtures of which the property studied depends on the proportions of the components but not their quantities in the mixture. A simplex is defined as a convex polyhedron with (k+1) vertices produced by k intersecting hyper planes in k-dimensional space. Any co-ordinate system above 3-dimensions are referred to as hyper planes, such planes are not orthogonal. A 2-dimensional regular simplex is, therefore, an equilateral triangle, while a 3-dimensional regular simplex is a regular tetrahedron. Scheffe used a regular (q -1) simplex to represent a factor space needed to describe a response surface for mixtures consisting of several components. If the number of components is denoted by q, then for binary system (q = 2) the required simplex is a straight line; for q = 3, the required simplex is an equilateral triangle; and for q = 4, the simplex is a regular tetrahedron. Strength and behavior of treated soft clay soils depend on the adequate proportioning of its ingredients or test materials. Scheffe developed an optimization theory that is used to optimize the behavior of treated soft soils. This considered experiments with mixtures or blends of which the property studied depends on the proportions or percentages by weight of the components but not their quantities in the mixture. He introduced polynomial regression to model the response, called "q, n -polynomials". These polynomials have to be of low degree (n), otherwise the polynomial contains a large number of coefficients, making interpretation difficult and requiring a large number of design points.

$$\text{if } n = 1: f(x) \sum_{i=1}^{q} \beta_i x_i \qquad (6.4.2)$$

$$\text{if } n = 2: f(x) \sum_{i=1}^{q} \beta_i x_i + \sum_{1 \leq i \leq j \leq q}^{q} \beta_i x_i x_j \qquad (6.4.3)$$

$$\text{if } n = 3: f(x) \sum_{i=1}^{q} \beta_i x_i + \sum_{1 \leq i \leq j \leq q}^{q} \beta_i x_i x_j + \sum_{1 \leq i \leq j \leq k \leq q}^{q} (\beta_{iij} x_i^2 x_j + \beta_{iij} x_i x_j x_k) \qquad (6.4.4)$$

6.4.2 Interaction of Compounds in Scheffe's Factor Space

Mixture designs are unique response surface designs where each of the components is bounded in values 0 and 1. Yet the sum of the components must be equal to one. If there are q components in a mixture, only q – 1 components are needed to determine the value of the last component. These constraints give rise to a (q – 1) dimensional design space called a simplex. That is, the dimension of the design space is always one less than the number of components. Thus the factor space is a regular (q – 1) dimensional simplex. In (q – 1) dimensional simplex if q = 2, we have 2 points of connectivity. This gives a straight line simplex lattice. If q=3, we have a triangular simplex lattice and for q = 4, it is however, a tetrahedron simplex lattice, etc. Taking a whole factor space in the design we have a (q, m) simplex lattice whose properties are defined as follows:

i. The factor space has uniformly distributed points,
ii. Simplex lattice designs are saturated

That is, the proportions used for each factor have m + 1 equally spaced levels from 0 to 1 (x_i = 0, 1/m, 2/m... 1), and all possible combinations are derived from such values of the component concentrations. This implies that all possible mixtures with these proportions are utilized. Hence, for the quadratic lattice (q, 2), approximating the response surface with the second degree polynomials (m=2), the following levels of every factor must be used 0, ½ and 1.

Number of Coefficients

$$P = 3, M = 2, N = \frac{(p+m-1)!}{m!(p-1)!} \text{ and } N = \frac{(3+2-1)!}{2!(3-1)!} = N = \frac{4!}{2!2!} = 6 \quad (6.4.5)$$

6.4.3 Three-Component Factor Space and Responses

The first three pseudo components are located at the vertices of the tetrahedron simplex; A_1 [1:0:0], A_2 [0:1:0], A_3 [0:0:1]. Three other pseudo mix ratios located at mid points of the lines joining the vertices of the simplex are A_{12} [0.5:0.5:0], A_{13} [0.5:0:0.5], A_{23} [0:0.5:0.5] as presented in Fig. 6.4.3a.

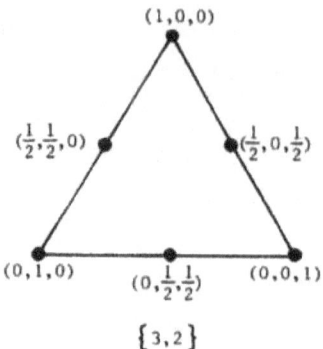

Fig. 6.4.3a Triangular simplex components

Responses are the selected properties of soil-additive blend or treatment matrix. A simplex lattice is described as a structural representation of lines joining the atoms of a mixture. The atoms are constituent components of the mixture. For the soil-additive blend mixture, the constituent elements are water, quarry dust and soil. And so it gives a simplex of a mixture of three components. Hence the simplex lattice of this three- component mixture is a three- dimensional solid equilateral triangle. Mixture components are subject to the constraint that the sum of all the components must be equal to one.

As a rule the response surfaces in multi-component systems are very intricate. To describe such surfaces adequately, high degree polynomials are required, and hence a great many experimental trials. A polynomial of degree n in q variable has C_{q+n}^n coefficients. If a mixture has a total of q components and X_i be the proportion of the ith component in the mixture such that,

$$X_i \geq 0 (i=1,2,...q) \qquad (6.4.6)$$

Then the sum of the component proportion is a whole unity i.e

$$X_1 + X_2 + X_3 = 1 \text{ or } \sum X_i - 1 = 0 \tag{6.4.7}$$

$$n = b_0 + \sum b_i X_i + \sum b_{ij} X_i X_j + \sum b_{ijk} X_i X_j X_k + ... + \sum b_{i1,i2}....in X_j, X_{i2}, X_{jn} \tag{6.4.8}$$

Where, $1 \leq i \leq q, 1 \leq j \leq q, 1 \leq i \leq j \leq q \leq k$ and $1 \leq i \leq j \leq k \leq l \leq .... \leq in \leq q$ respectively.

$$Y = b_0 + b_1 X_1 + b_2 X_2 + b_3 X_3 + b_{11} X_1^2 + b_{12} X_1 X_2 + b_{13} X_1 X_3 + b_{22} X_2^2 + b_{23} X_2 X_3 + b_{33} X_3^2 \tag{6.4.9}$$

Where, b is a constant coefficient.

The relationship obtainable from Eq. 6.4.9 is subjected to the normalization condition of Eq. 6.4.7 for a sum of independent variables. For a ternary mixture, which was dealt with in this work, the reduced second degree polynomial can be obtained as follows:

From Eq. 6.4.4

$$X_1 + X_2 + X_3 = 1 \tag{6.4.10}$$

i.e. $b_0 X_1 + b_0 X_2 + b_0 X_3 = b_0$ (6.4.11)

$b_0 = b_0 (X_1 + X_2 + X_3)$

Multiplying Eq. 6.4.10 by X_1, X_2, and X_3 in succession gives

$$\begin{aligned} X_1^2 &= X_1 - X_1 X_2 - X_1 X_3 \\ X_2^2 &= X_2 - X_1 X_2 - X_2 X_3 \\ X_3^2 &= X_3 - X_1 X_3 - X_2 X_3 \end{aligned} \tag{6.4.12}$$

Substituting Eq. 6.4.11 into Eq. 6.4.12, we obtain after necessary transformation that

$$\hat{Y} = (b_0+b_1+b_{11})X_1+(b_0+b_2+b_{22})X_2+(b_0+b_3+b_{33})X_3+(b_{12}+b_{11}+b_{22})X_1X_2$$
$$+(b_{13}+b_{11}+b_{33})X_1X_3+(b_{23}+b_{22}+b_{33})X_2X_3 \quad (6.4.13)$$

If we denote

$$\beta_i = b_0 + b_i + b_{ii}$$
$$\text{And } \beta_{ij} = b_{ij} - b_{ii} - b_{jj} \quad (6.4.14)$$

Then we arrive at the reduced second degree polynomial:

$$\hat{Y} = (b_0+b_1+b_{11})X_1+(b_0+b_2+b_{22})X_2+(b_0+b_3+b_{33})X_3+(b_{12}+b_{11}+b_{22})X_1X_2$$
$$+(b_{13}+b_{11}+b_{33})X_1X_3+(b_{23}+b_{22}+b_{33})X_2X_3 \quad (6.4.15)$$

$$\hat{Y} = (b_0+b_1+b_{11})X_1+(b_0+b_2+b_{22})X_2+(b_0+b_3+b_{33})X_3+(b_{12}+b_{11}+b_{22})X_1X_2$$
$$+(b_{13}+b_{11}+b_{33})X_1X_3+(b_{23}+b_{22}+b_{33})X_2X_3 \quad (6.4.16)$$

6.4.4 Actual Components and Pseudo Components

$$AZ = AX \quad (6.4.17)$$

Z represents the actual components while X represents the pseudo components, where A is the constant; a three by three matrix for the present work under study.

The value of matrix A will be obtained from the first three mix ratios. The mix ratios are;

Z_1 [0.10:0.2:1.0], Z_2 [0.15:0.5:1.0], Z_3 [0.25:0.95:1.0] (6.4.18)

The corresponding pseudo mix ratios are of an identity matrix form thus;

X_1 [1:0:0], X_2 [0: 1: 0], X_3 [0:0:1] (6.4.19)

Substitution of Xi and Zi into Eq. 6.4.16, then use the corresponding pseudo components to determine the corresponding actual mixture components.

X_1 = fraction of water ratio

X_2 = fraction of quarry dust

X_3 = fraction of soil

With the established mix proportions, the blend of mixtures are prepared for soils re-engineering operation.

Moreover, the sampling and proportioning of the test materials have been determined by a mix ratio model using the scheffe modelling mathematical method in this book. The proportions of the test materials; quarry dust and water were gotten from iterations of the 3, 2 scheffe polynomial. The values served as the percentage by weight of the dry solid added to the stabilization or treatment protocol. These exercises were conducted to validate the accuracy of the mathematical model. The specimens were prepared in that order and the swelling, California bearing ratio, unconfined compressive strength and durability tests also were experimented on under laboratory conditions.

6.4.5 Scheffe's Model Test for Adequacy and Validation

The test for adequacy of the model suggested by this book was done using Fischer test at 95% confidence level on the compressive strength at the control points i.e., C_1, C_2, C_3, C_{12}, C_{13} and C_{23}. In this test, two hypotheses were set as follows:

Null Hypothesis: There is no significant difference between the laboratory tests and model predicted swelling potential, California bearing ratio, compressive strength and durability results.

Alternative Hypothesis: There is a significant difference between the laboratory test and model predicted swelling potential, California bearing ratio, compressive strength and durability results.

Through the application of Scheffe's simplex model, the values of swelling potential, California bearing ratio, unconfined compressive strength and durability by loss of strength on immersion method of quarry dust treated soil were case studied and reported in this book. The model gave highest values of swelling potential of 6.2% corresponding to mix ratio of 0.1:0.2:1.0 for water, quarry dust, and soil respectively. The lowest response of swelling potential was found to be 2.4% corresponding to mix ratio of 0.25:0.95:1.0. This further showed that the improved value of swelling potential was achieved at a higher proportion of quarry dust, which was 95% by weight of solid The model gave highest values of California bearing ratio 32% corresponding to mix ratio of 0.25:0.95:1.0 for water, quarry dust, and soil respectively. The lowest response for California bearing ratio was found to be 15% corresponding to mix ratio of 0.1:0.2:1.0. This indicates that at higher proportion of quarry dust of 95%, the CBR of the treated soil recorded optimal value which meets the requirement of a materials utilization as a base material [68-74]. The model gave highest values of unconfined compressive strength of 258kN/m^2. This value corresponded to mix ratio of 0.25:0.95:1.0 for water, quarry dust, and soil respectively. The lowest response for unconfined compressive strength was found to be 90kN/m^2 corresponding to mix ratio of 0.1:0.2:1.0. The model gave highest values of loss of strength on immersion durability of 98%, which corresponded to mix ratio of 0.25:0.95:1.0 for water, quarry dust, and soil respectively. The lowest response for loss of strength on immersion durability was found to be 90% corresponding to mix ratio of 0.1:0.2:1.0. The maximum strength value was greater than the minimum value specified by the American association of state highway and transport officials (AASHTO) for the mechanical properties of soil. Using the model, swelling potential, California bearing ratio, unconfined compressive strength and loss of strength on immersion durability of all points in the simplex and perhaps other parameters of soils for a quarry dust treated soil can be derived. Taking a clue from the suggested operation in this book, other additive materials can be optimized by the scheffe's method for sustainable soils re-engineering. The variation index between laboratory and field situation may be negligible. This imploes that the suggested operation can be adapted for various situations and with many varied geomaterial additives.

6.5 Extreme Vertex Design for Sustainable Soils Re-Engineering

A mixture experiment is an experiment in which the response depends on the proportions of the components, not the total amount. There are two main constraints of mixture experiments. First, the proportion of a component is between 0 and 1. Second, the sum of proportions of all components is unity.

$$\sum_{i=1}^{q} x_i = 1 \quad (i = 1, 2, 3 .... q) \qquad (6.5.1)$$

Both constraints at the upper and lower regions affect the entire experimental region. The experimental region becomes a (q-1) simplex. An overview of the mixture experiment methodology was given by Cornell. Furthermore, additional constraints on proportions, such as lower bounds (L_b), upper bounds (U_b), (see Equation 6.5.2) will affect the shape of the experimental region.

$$0 \leq L_b \leq X_i \leq U_b \leq 1, (i = 1, 2, 3 .... q) \qquad (6.5.2)$$

The experimental region becomes a regular or irregular shape. Design points of the irregular shape of the mixture experiment of more than three components are difficult to determine by hand. It is needed for a computational approach. To determine the design points of an irregular mixture experiment is needed for a computational approach. Algorithms have been developed to select design points, planes, edges, vertices and centroids of experimental regions. One of such algorithms includes the XVERT algorithm developed to find the design points in the linear model by Snee. The XVERT algorithm can be used for selecting a subset of extreme vertices when the number of candidate vertices is large. The linear model can be described by Scheffe as shown in Equation 6.5.3

$$\hat{y}(x) = \sum_{i=1}^{q} Y_1 x_i \tag{6.5.3}$$

Subsequently, the XVERT algorithm to find the design points in the quadratic model was developed by Snee. The mixture design for a quadratic model produces large experimental runs. The centroids are calculated by averaging various subsets of vertices. The quadratic model Scheffe can be described by

$$\hat{y}(x) = \sum_{i=1}^{q} Y_i x_i \sum_{i=1}^{q-1} \sum_{j=i+1}^{q} Y_{ij} x_i x_j \tag{6.5.4}$$

And,

$$p = \frac{q(q+1)}{2} \tag{6.5.5}$$

Eq. 5 is the number of parameter in the quadratic model.

A design which minimizes the determinant of variance (y) or maximizes the determinant of the information matrix [M] is called D-optimal design. The D-optimality criterion is defined as

$$D = \max |M| = \max |X^T X| \tag{6.5.6}$$

Algorithms start with writing 2^{q-1} combinations of upper and lower bounds for all but one factor which is left blank as in Mclean. The extreme vertices also can be computed using the XVERT Algorithm steps and sequences describe below;

- Rank the components in order of increasing $U_i - L_i$, X_1 ranges has the smallest range and X_q has the largest range.
- Consider first components q-1 with the smallest ranges. Form a two-level design from the lower - upper bounds of these q-1 components. There are 2^{q-1} combinations.
- Determine the level of the omitted component X_q with each of the 2^{q-1} combination in step 2 using $X_q = 1 - \sum_{i=1}^{q} x_i$

- If this computed value lie within the constraint limits it is an extreme vertex called as core point. If it falls outside the constraint limits of the corresponding component it is called as the candidate point. For the points which are outside of the constraint limits, set X_q equal to the upper or lower limit, whichever is closest to the computed value
- Additional points are generated from the candidate points. Find the difference between computed value and substituted upper or lower limit. Adjust this difference to one of the q-1 components. The generated point is an extreme vertex if the level after adjustment remains within the limits of the components. Thus maximum q-1 points can be generated from one candidate point.

In general, extreme vertices method has been used in various fields of science of experimentation and mixture blending and more prominent in this effort is the medical sciences. Recently it has been adopted in the production and blending of cementing materials like the geopolymer cement. This work has targeted adapting this method in various fields of design of experiments in civil engineering which include; concrete production and modification, soil stabilization, asphalt production, and water treatment. To accomplish these tasks in civil engineering, components are blended in proportion utilizing both primary and secondary components depending on the conditions of the blending. Four technical cases were reviewed in this work; (i) a 5- component experimental mixture for concrete production utilizing water proportion, cement proportion, palm bunch ash proportion, fine aggregate proportion, and coarse aggregate proportion. The blending of components form an experimental space called the simplex as shown in Fig. 1. This forms the space within which the behavior of the homogenous blend resulting from the mixing of the experimental components are distributed.

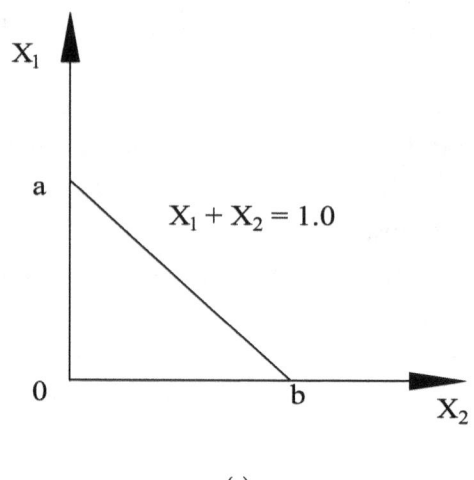

(a)

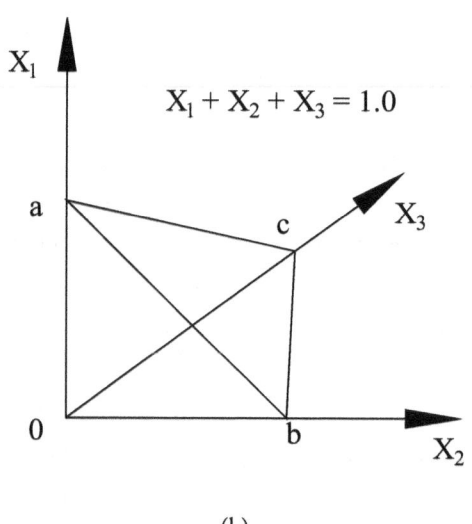

(b)

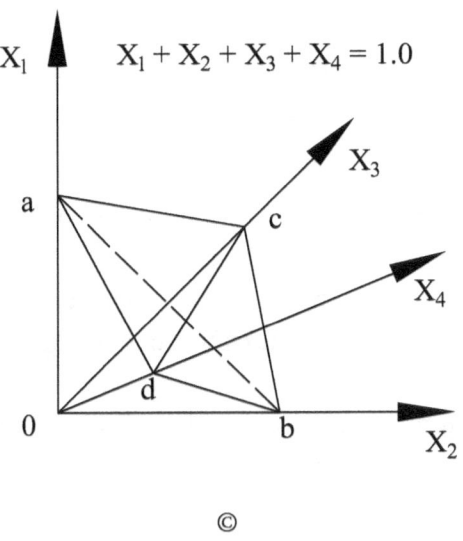

(c)

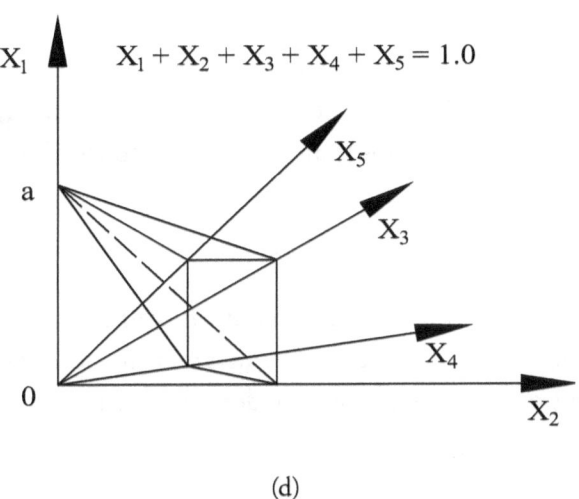

(d)

Fig. 6.5.1 Extreme vertices for; (a) 2-component simplex, 3- component simplex, 4- component simplex and 5- component simplex

6.5.5 Formulation of Constraints and Design of Factor Space

Constraints Formulation

Constraints are regions of lower and upper bounds established by the properties of the components that make up an experimental blend. As soon as these components are decided on based on intended results, the constraints that would define the experimental region are selected from available resources. In most cases and in practice, physical and economic considerations impose most often the lower and upper limits. Snee had proposed general constraints equation as follows;

$$0 \leq L_i \leq X_i \leq U_i \ 1 \ (i = 1, 2, 3 \ldots q) \tag{6.5.1.1}$$

Where; L_i equals lower bound, U_i equals upper bound, X_i equals the ith component and q is the number of components in the mixture. Snee also suggested an equation for multiple variable constraints for the form;

$$C_j \leq A_{1j}X_1 + A_{2j}X_2 + \ldots + A_{qj}X_q \leq D_j \tag{6.5.1.2}$$

, which are also found in experimentation and design of mixture where $C_j = D_j$ for all j = 1, 2, 3 ..., m are scalar constants specified by multicomponent mixture and j designate the minor component proportion.

A consideration of some selected cases found in practice in various civil engineering disciplines are discussed as follows;

Case 1: Constraints of a five (5) component experimental mixture for concrete production: the multicomponent constraints in Eqns. 9-14 have been developed from concrete production literature references and end conditions from earlier research results on the utilization of additives as partial replacement for ordinary cement or as an enhancer of concrete mixes in concrete production. Under the conditions of an additive serving as partial replacement for cement with cementing or pozzolanic properties, it is considered a minor component in a mixture of mixture experiment (MME).

$$0.6 \leq X_4 + X_5 \leq 0.75 \quad (6.5.1.3)$$

$$0.1 \leq X_2 + X_3 \leq 0.35 \quad (6.5.1.4)$$

$$0.1 \leq X_1 \leq 0.15 \quad (6.5.1.5)$$

$$0.45 \leq \frac{X_1}{X_2 + X_3} \leq 0.55 \quad (6.5.1.6)$$

$$0.05 \leq \frac{X_3}{X_2} \leq 0.25 \quad (6.5.1.7)$$

$$X_1 + X_2 + X_3 + X_4 + X_5 = 1 \quad (6.5.1.8)$$

Where; X_1 equals water proportion, X_2 equals cement proportion, X_3 equals palm bunch ash proportion, X_4 equals fine aggregate proportion, and X_5 equals coarse aggregate proportion.

Case 2: Constraints of a four (4) component experimental mixture for asphalt production: in a similar operation the multicomponent constraints in Eqns. 6.5.1.9 - 6.5.1.12 have been developed from asphalt production and modification literature references and end conditions from research results on the utilization of crushed waste glasses based geopolymer cement as a modifier. In this case, the modifier is a proportion of the major cementing material in asphalt production i.e. the asphalt cement particularly shown in Eq. 6.5.1.12.

$$1.01 \leq X_1 \leq 0.05 \quad (6.5.1.9)$$

$$0.75 \leq X_2 + X_3 \leq 0.95 \quad (6.5.1.10)$$

$$0.15 \leq \frac{X_4}{X_1} \leq 0.45 \quad (6.5.1.11)$$

$$X_1 + X_2 + X_3 + X_4 = 1.0 \quad (6.5.1.12)$$

Case 3: Constraints of a three (3) component experimental mixture for soil treatment: in soil stabilization protocols, materials are blended with the treated soil to improve on its engineering properties. The utilization of quarry dust as an admixture has been in use in various circumstances and reported in many literatures. The results achieved from the above operation have been helpful in the formulation of the multicomponent constraints as in Eqns. 6.5.1.13-6.5.1.15.

$$1.1 \leq \frac{X_1}{X_3} \leq 0.9 \qquad (6.5.1.13)$$

$$0.1 \leq X_2 \leq 0.15 \qquad (6.5.1.14)$$

$$X_1 + X_2 + X_3 = 1.0 \qquad (6.5.1.15)$$

Where; X_1 equals quarry dust proportion, X_2 equals water content and X_3 equals test soil proportion.

Case 4: Constraints of a two (2) component experimental mixture of homogenous blend for example the improvement of freshly mixed concrete properties with freshly synthesized quarry dust based geopolymer cement. In a similar way, the constraints as in Eqns. 6.5.1.16-6.5.1.18 have been proposed from earlier research works. It is important to also note that the synthesized quarry dust based geopolymer cement functions as a minor component in a partial replacement technique for the concrete or another case could serve as an additive in a side by side utilization as a major component for the improvement of certain properties in concrete for example durability, heat resistance, sulfate resistance, shrinkage resistance and cracking resistance.

$$X_1 \leq 1.0 \qquad (6.5.1.16)$$

$$0.1 \leq \frac{X_1}{X_2} \leq 0.55 \qquad (6.5.1.17)$$

$$X_1 + X_2 = 1.0 \qquad (6.5.1.18)$$

Where; X_1 equals the homogenous freshly mixed concrete proportion and X_2 equals the homogenous freshly synthesized geopolymer cement.

6.5.6 Design of Simplex and Factor Space

5- Component Simplex and Factor Space for Concrete Production

The design of factor spaces from hyper-polyhedron simplexes begins with the testing and screening of the components constraints giving rise to an experimental points within the defined or constrained space. In the case of the 5- component factor space under review considerations, multicomponent constraints were developed from literatures on concrete production and modification. These constraints were used to test and evaluate the degrees of freedom (df) in the 5 factors component design experiment shown in Table 6.5.5.1. A recommendation is a minimum of 3 lack of fit df and 4 df for pure error. This ensures a valid lack of fit test. Fewer df will lead to a test that may not detect lack of fit.

Table 6.5.5.1. Design Matrix Evaluation for Mixture Quadratic Model 5 Factors: A, B, C, D, E

Mixture Component Coding is U_Pseudo.
Degrees of Freedom for Evaluation

Model	14
Residuals	10
Lack of Fit	5
Pure Error	5
Corr Total	24

Power calculations test was also conducted on the developed constraints using the design expert and the Minitab software to establish the deviations and variances on the design planes and vertexes and edges contained in the simplex on 5% alpha level shown in Table 6.5.5.2;

Table 6.5.5.2. Power at 5 % alpha level on 5- component for concrete production

Term	StdErr	VIF	Ri-Squared	Std. Dev.
A	8.18	80.41	0.9876	5.5 %

B	1.50	7.50	0.8666	9.8 %
C	6.52	62.76	0.9841	5.7 %
D	2.41	14.50	0.9311	8.8 %
E	0.70	1.82	0.4503	10.3 %
AB	14.27	22.58	0.9557	8.0 %
AC	17.28	15.72	0.9364	7.0 %
AD	14.76	19.62	0.9490	7.8 %
AE	14.31	16.33	0.9388	8.0 %
BC	11.28	16.25	0.9385	9.9 %
BD	6.73	4.40	0.7725	19.0 %
BE	4.13	2.36	0.5759	41.8 %
CD	11.82	13.97	0.9284	9.4 %
CE	12.43	13.01	0.9232	9.0 %
DE	5.83	3.05	0.6726	23.7 %

Basis Std. Dev. = 1.0

Approximate DF used for power calculations operate under the following condition;

- Standard errors should be similar within type of coefficient. Smaller is better.
- The ideal VIF value is 1.0. VIFs above 10 are cause for concern. VIFs above 100 are cause for alarm, indicating coefficients are poorly estimated due to multicollinearity.
- Ideal Ri-squared is 0.0. High Ri-squared means terms are correlated with each other, possibly leading to poor models.
- For mixture designs the proportions of components must sum to one.
- This is a constraint on the system and causes multicollinearity to exist, thus increasing the VIFs and the Ri-squareds, rendering these statistics useless.

The software further developed the conditions of the 5- component simplex shown in Fig. 6.5.2 and the results are presented in Table 6.5.5.3. The 25 runs were to improve on the optimality or efficiency of the model operation. Lack of fit was never recorded on any of the vertex points of the design space as shown in Table 3 rather on either the interior or plane points. This in effect raises concern for more design points to be located on the interior and plane spaces of the simplex to reduce the lack of fit effect on the experimental space.

Table 6.5.5.3. Measures derived from the information matrix on 5- component for concrete production

Run	Leverage	Space Type	Build Type
1	0.2731	Interior	Lack of Fit
2	0.8503	Edge	Model
3	0.2550	Plane	Replicate
4	0.4124	Plane	Lack of Fit
5	0.2550	Plane	Lack of Fit
6	0.4771	Edge	Replicate
7	0.7860	Edge	Model
8	0.3999	Plane	Model
9	0.3999	Plane	Replicate
10	0.3989	Plane	Replicate
11	0.8193	Vertex	Model
12	0.9334	Vertex	Model
13	0.8727	Vertex	Model
14	0.4901	Vertex	Model
15	0.8335	Vertex	Model
16	0.4901	Vertex	Replicate
17	0.8508	Vertex	Model
18	0.4175	Interior	Lack of Fit
19	0.7665	Edge	Model
20	0.8631	Vertex	Model
21	0.3989	Plane	Model
22	0.8293	Vertex	Model
23	0.9410	Vertex	Model
24	0.4771	Edge	Model
25	0.5091	Plane	Lack of Fit
Average =	0.6000		

However, watch for leverages close to 1.0 because they appear to be located on the vertexes and edges and consider replicating these points or make sure they are run very carefully. The software generates lots of other data that would be used to test the multicollinearity of the design, the G-efficiency and the scaled D- optimality. These information and results are needed when comparing designs.

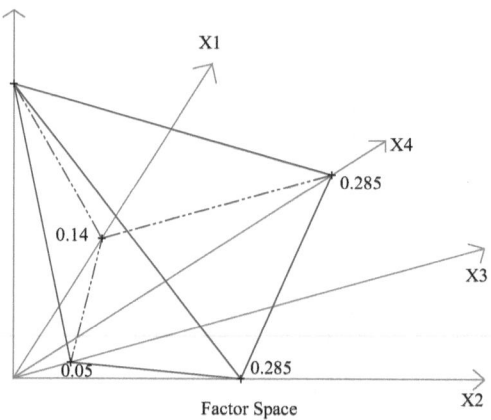

Fig. 6.5.2. Factor space simplex of a 5- component mixture experiment for concrete production

4- Component Simplex and Factor Space

Table 6.5.5.4 shows the design evaluation for the four component mixture quadratic model conducted with the multicomponent constraints developed from literature to determine the degree of freedom for the experimental procedure of an asphalt production and modification exercise. A recommendation is a minimum of 3 lack of fit df and 4 df for pure error. This ensures a valid lack of fit test. Fewer df will lead to a test that may not detect lack of fit.

Table 6.5.5.4. Design Matrix Evaluation for Mixture Quadratic Model 4 Factors: A, B, C, D with U_Pseudo Mixture Component Coding;

Degrees of Freedom for Evaluation
    Model   9

Residuals	15
Lack of Fit	8
Pure Error	7
Corr Total	24

Power calculations test was also conducted on the developed constraints using the design expert and minitab software to find the standard deviations and variances on the design planes and vertexes and edges contained in the simplex on 5% alpha level shown in Table 6.5.5.5.

Table 6.5.5.5. Power at 5 % alpha level on 4- component for asphalt production

Term	StdErr	VIF	Ri-Squared	Std. Dev.
A	14.56	208.97	0.9952	5.1 %
B	1.42	9.63	0.8961	5.4 %
C	0.62	2.52	0.6034	5.3 %
D	34.51	510.91	0.9980	5.0 %
AB	21.04	64.80	0.9846	6.5 %
AC	20.97	73.79	0.9864	6.5 %
AD	43.66	35.04	0.9715	5.3 %
BC	3.38	2.76	0.6377	60.1 %
BD	44.85	145.55	0.9931	5.3 %
CD	43.28	132.19	0.9924	5.3 %

Basis Std. Dev. = 1.0

Approximate DF used for power calculations functions under the following;

- Standard errors should be similar within type of coefficient. Smaller is better.
- The ideal VIF value is 1.0. VIFs above 10 are cause for concern. VIFs above 100 are cause for alarm, indicating coefficients are poorly estimated due to multicollinearity.
- Ideal Ri-squared is 0.0. High Ri-squared means terms are correlated with each other, possibly leading to poor models.

- For mixture designs the proportions of components must sum to one.
- This is a constraint on the system and causes multicollinearity to exist, thus increasing the VIFs and the Ri-squareds, rendering these statistics useless.

The software further developed the conditions of the 4- component simplex shown in Figs. 6.5.4 & 6.5.5 and the results are presented in Table 6.5.5.6. The 25 runs were to improve on the optimality or efficiency of the model operation. Lack of fit was recorded on one vertex point of the design space in this case as shown in Table 6 and on the third edge and axial points. This in effect raises concern for more design points to be located on the third edge and axial spaces of the simplex to reduce the lack of fit effect on the experimental space.

Table 6.5.5.6. Measures derived from the information matrix on 4- component for asphalt production

Run	Leverage	Space Type	Build Type
1	0.3356	ThirdEdge	Lack of Fit
2	0.1901	Center	Center
3	0.3344	ThirdEdge	Replicate
4	0.5196	Vertex	Model
5	0.1901	Center	Center
6	0.3344	ThirdEdge	Model
7	0.4232	ThirdEdge	Model
8	0.1901	Center	Center
9	0.3225	ThirdEdge	Model
10	0.4148	CentEdge	Model
11	0.83257	Vertex	Model
12	0.1747	AxialCB	Lack of Fit
13	0.4417	Vertex	Lack of Fit
14	0.3368	TripBlend	Model
15	0.3884	Vertex	Replicate
16	0.5385	Vertex	Model

17	0.3884	Vertex	Model
18	0.7909	Vertex	Model
19	0.3030	PlaneCent	Model
20	0.3562	ThirdEdge	Lack of Fit
21	0.4232	ThirdEdge	Replicate
22	0.3030	PlaneCent	Replicate
23	0.3241	ThirdEdge	Lack of Fit
24	0.3368	TripBlend	Replicate
25	0.807231	Vertex	Model
Average =	0.4000		

However, watch for leverages close to 1.0 because they appear to be located on the vertexes and edges and consider replicating these points or make sure they are run very carefully. The software generates lots of other data that would be used to test the multicollinearity of the design, the G-efficiency and the scaled D- optimality. These information and results are needed when comparing designs.

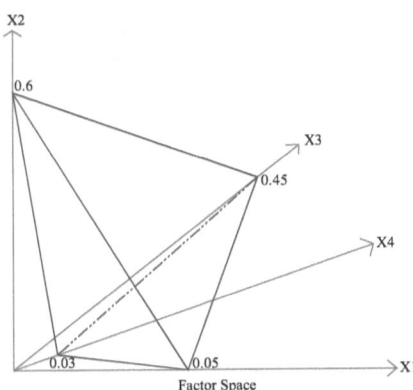

Fig. 6.5.4. Factor space simplex of a 4- component mixture experiment for asphalt production

3- Component Simplex and Factor Space

The design evaluation for the three component mixture quadratic model conducted with the multicomponent constraints developed from literature to determine the degree of freedom for the experimental procedure of a soil stabilization protocol is as presented in Table 6.5.5.7. A recommendation is a minimum of 3 lack of fit df and 4 df for pure error. This ensures a valid lack of fit test. Fewer df will lead to a test that may not detect lack of fit.

Table 6.5.5.7. Design Matrix Evaluation for Mixture Quadratic Model 3 Factors: A, B, C with L_Pseudo Mixture Component Coding

Degrees of Freedom for Evaluation

Model	5
Residuals	19
Lack of Fit	7
Pure Error	12
Corr Total	24

Power calculations test was also conducted on the developed constraints using the design expert and minitab software to find the standard deviations and variances on the design planes and vertexes and edges contained in the simplex on 5% alpha level shown in Table 8

Table 8. Power at 5 % alpha level on 3- component for soil treatment

Term	StdErr	VIF	Ri-Squared	Std. Dev.
A	0.52	2.42	0.5860	6.4 %
B	11.15	131.86	0.9924	5.3 %
C	1.81	13.14	0.9239	6.1 %
AB	16.13	64.60	0.9845	7.6 %
AC	4.11	7.48	0.8663	45.5 %
BC	16.55	45.85	0.9782	7.5 %

Basis Std. Dev. = 1.0

Approximate DF used for power calculations functions as follows;

- Standard errors should be similar within type of coefficient. Smaller is better.

- The ideal VIF value is 1.0. VIFs above 10 are cause for concern. VIFs above 100 are cause for alarm, indicating coefficients are poorly estimated due to multicollinearity.
- Ideal Ri-squared is 0.0. High Ri-squared means terms are correlated with each other, possibly leading to poor models.
- For mixture designs the proportions of components must sum to one.
- This is a constraint on the system and causes multicollinearity to exist, thus increasing the VIFs and the Ri-squareds, rendering these statistics useless.

The software further developed the conditions of the 3- component simplex shown in Figs. 6.5.6 & 6.5.7 and the results are presented in Table 6.5.5.9. The 25 runs were to improve on the optimality or efficiency of the model operation. Lack of fit was recorded on three interior points of the design space in this case as shown in Table 6.5.5.9 and on two edge points. This in effect raises concern for more design points to be located on these spaces of the simplex to reduce the lack of fit effect on the entire experimental space.

Table 6.5.5.9. Measures derived from the information matrix on 3- component for soil treatment

Run	Leverage	Space Type	Build Type
1	0.1200	Interior	Lack of Fit
2	0.3614	Vertex	Model
3	0.1314	Center	Center
4	0.2745	Vertex	Model
5	0.2377	Edge	Model
6	0.2477	Edge	Model
7	0.3614	Vertex	Model
8	0.1185	Interior	Lack of Fit
9	0.3364	Vertex	Model
10	0.2745	Vertex	Replicate
11	0.4050	Vertex	Model

12	0.1314	Center	Center
13	0.2745	Vertex	Model
14	0.3364	Vertex	Replicate
15	0.1314	Center	Center
16	0.2460	Edge	Replicate
17	0.2477	Edge	Replicate
18	0.2418	Edge	Replicate
19	0.2418	Edge	Model
20	0.1332	Interior	Lack of Fit
21	0.2460	Edge	Lack of Fit
22	0.1314	Center	Center
23	0.1314	Center	Center
24	0.2335	Edge	Lack of Fit
25	0.4050	Vertex	Model
Average =	0.2400		

However, watch for leverages close to 1.0 because they appear to be located on none of the design points in this case. The software generates lots of other data that would be used to test the multicollinearity of the design, the G-efficiency and the scaled D- optimality. These information and results are needed when comparing design.

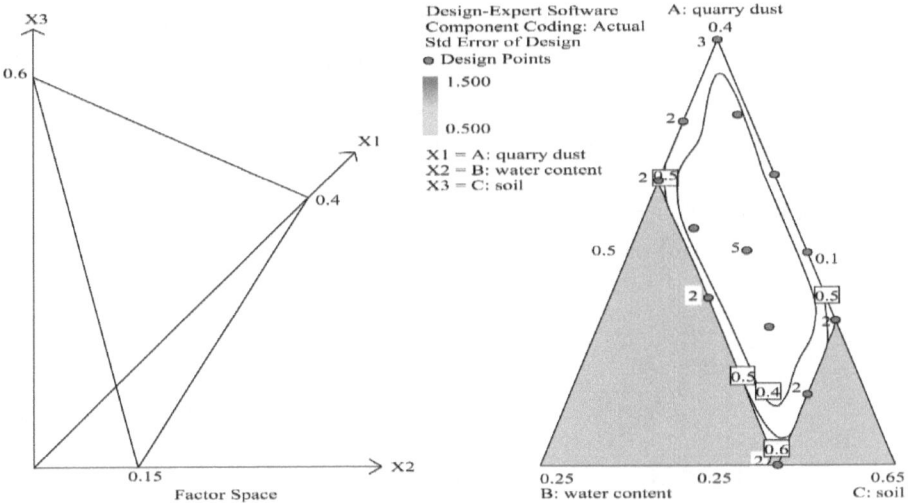

Fig. 6.5.6. Factor space simplex and contour space of a 3- component mixture experiment for soil stabilization

2- Component Simplex and Factor Space

In the final case scenario being reviewed, the design evaluation for the two component mixture quadratic model conducted with the multicomponent constraints developed from literature to determine the degree of freedom for the experimental procedure of a two homogenous mixture concrete modification protocol is as presented in Table 6.5.5.10. As usual, a recommendation is a minimum of 3 lack of fit df and 4 df for pure error. This ensures a valid lack of fit test. Fewer df will lead to a test that may not detect lack of fit.

Table 6.5.5.10. Design Matrix Evaluation for Mixture Quadratic Model 2 Factors: A, B with L_Pseudo Mixture Component Coding;

Degrees of Freedom for Evaluation
- Model   2
- Residuals   4
- *Lack of Fit*   4
- *Pure Error*   0
- Corr Total   6

Power calculations test was also conducted on the developed constraints using the design expert and minitab software to find the standard deviations and variances on the design planes and vertexes and edges contained in the simplex on 5% alpha level shown in Table 6.5.5.11.

Table 6.5.5.11. Power at 5 % alpha level on 2-component for homogeneous mixtures

Term	StdErr	VIF	Ri-Squared	Std. Dev.
A	0.92	2.07	0.5174	24.7 %
B	0.92	2.07	0.5174	24.7 %
AB	3.83	3.40	0.7055	36.0 %

Basis Std. Dev. = 1.0

Approximate DF used for power calculations.

- Standard errors should be similar within type of coefficient. Smaller is better.
- The ideal VIF value is 1.0. VIFs above 10 are cause for concern. VIFs above 100 are cause for alarm, indicating coefficients are poorly estimated due to multicollinearity.
- Ideal Ri-squared is 0.0. High Ri-squared means terms are correlated with each other, possibly leading to poor models.
- For mixture designs the proportions of components must sum to one.
- This is a constraint on the system and causes multicollinearity to exist, thus increasing the VIFs and the Ri-squareds, rendering these statistics useless.

The software further also developed the conditions of the 2- component simplex shown in Fig. 6.5.8 and the results are presented in Table 6.5.5.12. The 7 runs were to improve on the optimality or efficiency of the model operation. Lack of fit was not recorded on any of the design points.

Table 6.5.5.12. Measures derived from the information matrix on 2- component

Run	Leverage	Space Type
1	0.2815	Center
2	0.8525	Vertex
3	0.2524	AxialCB
4	0.2544	ThirdEdge
5	0.2524	AxialCB
6	0.2544	ThirdEdge
7	0.8525	Vertex
Average =	0.4286	

Watch for leverages close to 1.0. Consider replicating these points or make sure they are run very carefully. This case was observed on the 7th run located on the vertex of the experimental space. The software generates lots of other data that would be used to test the multicollinearity of the design, the G-efficiency and the scaled D- optimality. These information and results are needed when comparing designs.

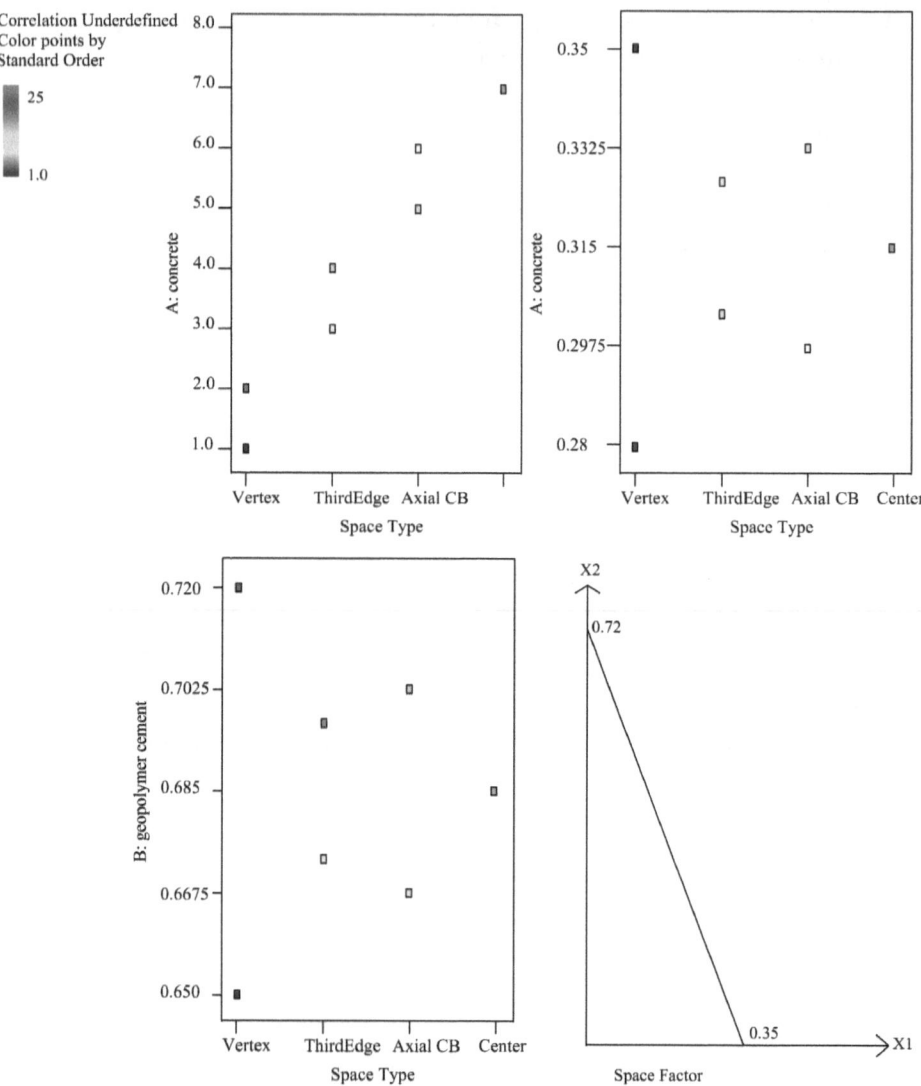

Fig. 6.5.8. Experimental simplex and factor space of the components in a 2-component mixture space

SUSTAINABLE SOILS RE-ENGINEERING

6.5.7 Design of Experimental Mix Proportions

Tables 6.5.7.13, 6.5.7.14, 6.5.7.15 and 6.5.7.16 present the mixes and runs for the 5-, 4-, 3-, and 2- component multiconstraints experimental design. These mixes guide the preparation of specimens to be tested in the laboratory to achieve the responses. The number of runs can be increased to check and screen for errors and reduce lack of fit effects within the experimental space. The specimens are prepared with the actual components mix proportions of the different components that make the test blend. Figs. 6.5.9-6.5.15 show the factor spaces, traces and deviations and contour of the different multicomponent constraints mixture of mixture experiments. It would be appropriate that in a model exercise, the full simulation of the behavior of the tested specimens are observed and shown graphically to enable engineers monitor the performance and life service of such infrastructures. These designed mixes would guide from experimental stage to achieve laboratory responses that enable the establishment of model equations that would determine the overall behavior of the modeled facility. Experimental responses are key to validating and testing the accuracy of mathematical modeling exercise as this under review. This research is confident that it serves as a hub to direct and guide exercises in the field of civil engineering in adapting extreme vertex design in all mixture experimental and composite formulations in civil engineering and even in industrial and materials mechanical engineering.

Table 6.5.7.13. 5- Component experimental mix proportions

Runs	Actual Components					Response	Pseudo Components				
	z_1	z_2	z_3	z_4	z_5		x_1	x_2	x_3	x_4	x_5
1	0.132	0.177	0.010	0.284	0.396	Y1	0.080	0.246	0.417	0.009	0.248
2	0.124	0.200	0.005	0.251	0.420	Y2	0.170	0.000	0.474	0.356	0.000
3	0.106	0.175	0.039	0.285	0.395	Y3	0.361	0.261	0.113	0.000	0.265
4	0.124	0.200	0.047	0.231	0.398	Y4	0.171	0.000	0.031	0.571	0.227
5	0.106	0.175	0.039	0.285	0.395	Y5	0.361	0.261	0.113	0.000	0.265
6	0.140	0.152	0.050	0.285	0.373	Y6	0.000	0.507	0.000	0.000	0.493

7	0.130	0.180	0.050	0.220	0.420	Y7	0.104	0.212	0.000	0.684	0.000
8	0.126	0.139	0.030	0.285	0.420	Y8	0.150	0.643	0.207	0.000	0.000
9	0.126	0.139	0.030	0.285	0.420	Y9	0.150	0.643	0.207	0.000	0.000
10	0.140	0.163	0.029	0.249	0.420	Y10	0.000	0.394	0.223	0.384	0.000
11	0.140	0.200	0.005	0.285	0.370	Y11	0.000	0.000	0.474	0.000	0.526
12	0.100	0.200	0.050	0.230	0.420	Y12	0.421	0.000	0.000	0.579	0.000
13	0.140	0.120	0.050	0.270	0.420	Y13	0.000	0.842	0.000	0.158	0.000
14	0.140	0.200	0.050	0.285	0.325	Y14	0.000	0.000	0.000	0.000	1.000
15	0.100	0.145	0.050	0.285	0.420	Y15	0.421	0.579	0.000	0.000	0.000
16	0.140	0.200	0.050	0.285	0.325	Y16	0.000	0.000	0.000	0.000	1.000
17	0.100	0.200	0.050	0.285	0.365	Y17	0.421	0.000	0.000	0.000	0.579
18	0.140	0.195	0.021	0.254	0.390	Y18	0.001	0.050	0.307	0.324	0.318
19	0.140	0.200	0.050	0.238	0.372	Y19	0.000	0.000	0.000	0.492	0.508
20	0.140	0.150	0.005	0.285	0.420	Y20	0.000	0.526	0.474	0.000	0.000
21	0.140	0.163	0.029	0.249	0.420	Y21	0.000	0.394	0.223	0.384	0.000
22	0.140	0.200	0.020	0.220	0.420	Y22	0.000	0.000	0.316	0.684	0.000
23	0.100	0.190	0.005	0.285	0.420	Y23	0.421	0.105	0.474	0.000	0.000
24	0.140	0.152	0.050	0.285	0.373	Y24	0.000	0.507	0.000	0.000	0.493
25	0.109	0.168	0.050	0.253	0.420	Y25	0.331	0.333	0.000	0.336	0.000

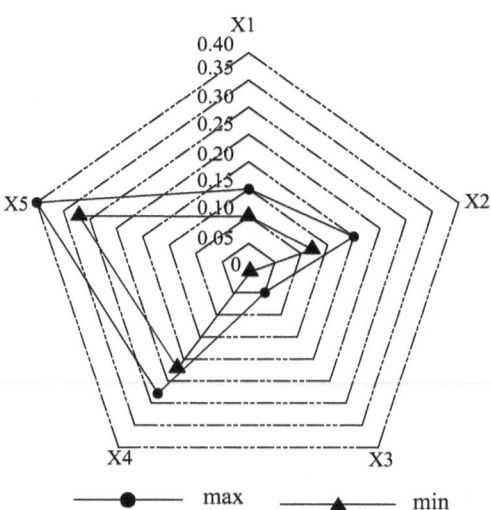

Fig. 6.5.9. Array factor space of the 5- component simplex of concrete production

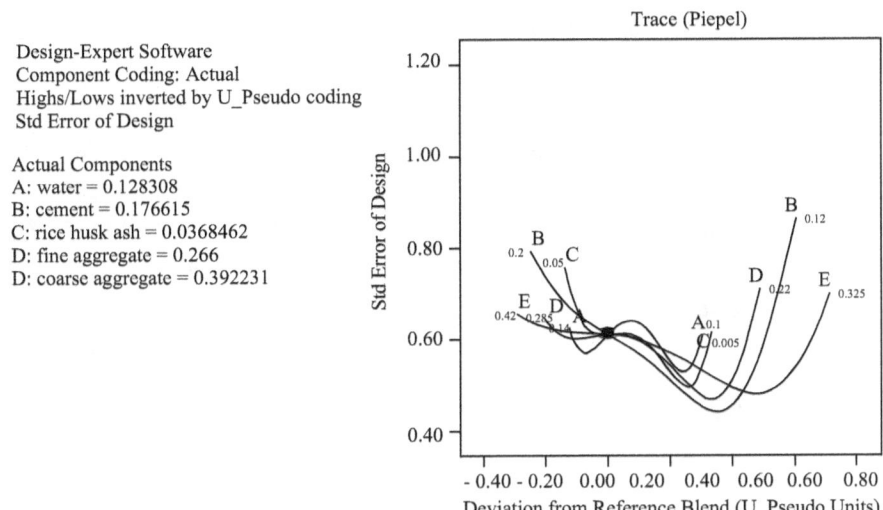

Fig. 6.5.10. Trace and deviation factor space of the 5- component mixture for concrete production

Table 6.5.7.14. 4- Component experimental mix proportion

Runs	Actual Components				Response	Pseudo Components			
	z1	z2	z3	z4		x1	x2	x3	x4
1	0.010	0.540	0.420	0.030	Y1	0.308	0.462	0.231	0.000
2	0.031	0.545	0.409	0.016	Y2	0.147	0.424	0.319	0.110
3	0.030	0.500	0.450	0.021	Y3	0.158	0.769	0.000	0.073
4	0.010	0.539	0.450	0.002	Y4	0.308	0.473	0.000	0.219
5	0.031	0.545	0.409	0.016	Y5	0.147	0.424	0.319	0.110
6	0.030	0.500	0.450	0.021	Y6	0.158	0.769	0.000	0.073
7	0.010	0.600	0.379	0.011	Y7	0.308	0.000	0.546	0.146
8	0.031	0.545	0.409	0.016	Y8	0.147	0.424	0.319	0.110
9	0.010	0.570	0.390	0.030	Y9	0.308	0.231	0.462	0.000
10	0.030	0.600	0.340	0.030	Y10	0.154	0.000	0.846	0.000
11	0.050	0.500	0.420	0.030	Y11	0.000	0.769	0.231	0.000
12	0.040	0.522	0.429	0.009	Y12	0.074	0.597	0.165	0.164
13	0.010	0.600	0.360	0.030	Y13	0.308	0.000	0.692	0.000
14	0.023	0.580	0.396	0.002	Y14	0.205	0.158	0.418	0.219
15	0.050	0.600	0.320	0.030	Y15	0.000	0.000	1.000	0.000

16	0.010	0.510	0.450	0.030	Y16	0.308	0.692	0.000	0.000
17	0.050	0.600	0.320	0.030	Y17	0.000	0.000	1.000	0.000
18	0.050	0.600	0.349	0.002	Y18	0.000	0.000	0.781	0.219
19	0.050	0.550	0.384	0.016	Y19	0.000	0.385	0.506	0.110
20	0.010	0.559	0.430	0.002	Y20	0.308	0.315	0.158	0.219
21	0.010	0.600	0.379	0.011	Y21	0.308	0.000	0.546	0.146
22	0.050	0.550	0.384	0.016	Y22	0.000	0.385	0.506	0.110
23	0.050	0.567	0.353	0.030	Y23	0.000	0.256	0.744	0.000
24	0.023	0.580	0.396	0.002	Y24	0.205	0.158	0.418	0.219
25	0.050	0.500	0.449	0.002	Y25	0.000	0.769	0.012	0.219

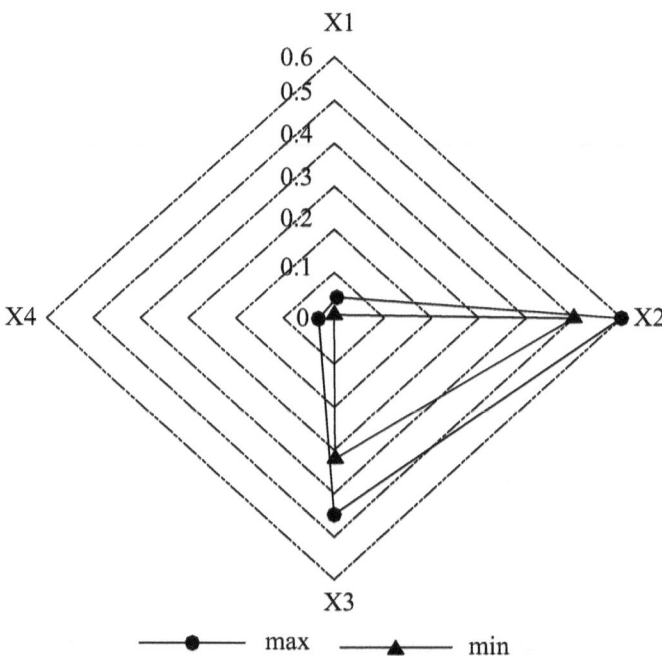

Fig. 6.5.11. Array factor space of the 4- component simplex of asphalt production

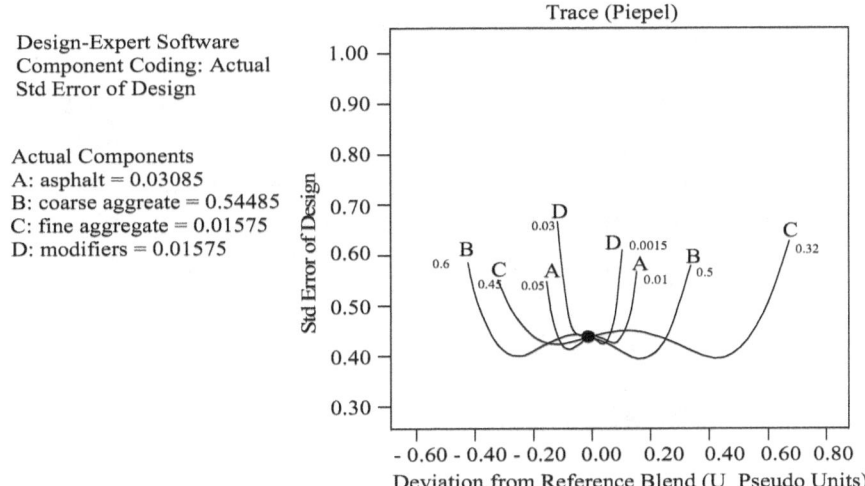

Fig. 6.5.12. Trace and deviation factor space of the 4- component mixture for asphalt production

Table 6.5.7.15. 3- Component experimental mix proportions for soil stabilization

Runs	Actual Components			Response	Pseudo Components		
	z1	z2	z3		x1	x2	x3
1	0.333	0.143	0.524	Y1	0.551	0.289	0.159
2	0.300	0.100	0.600	Y2	0.333	0.000	0.667
3	0.325	0.125	0.550	Y3	0.500	0.167	0.333
4	0.400	0.100	0.500	Y4	1.000	0.000	0.000
5	0.352	0.100	0.548	Y5	0.679	0.000	0.321
6	0.371	0.129	0.500	Y6	0.804	0.196	0.000
7	0.300	0.100	0.600	Y7	0.333	0.000	0.667
8	0.298	0.129	0.573	Y8	0.322	0.193	0.485
9	0.350	0.150	0.500	Y9	0.667	0.333	0.000
10	0.400	0.100	0.500	Y10	1.000	0.000	0.000
11	0.250	0.150	0.600	Y11	0.000	0.333	0.667
12	0.325	0.125	0.550	Y12	0.500	0.167	0.333
13	0.400	0.100	0.500	Y13	1.000	0.000	0.000
14	0.350	0.150	0.500	Y14	0.667	0.333	0.000

15	0.325	0.125	0.550	Y15	0.500	0.167	0.333
16	0.275	0.125	0.600	Y16	0.165	0.169	0.667
17	0.371	0.129	0.500	Y17	0.804	0.196	0.000
18	0.308	0.150	0.542	Y18	0.390	0.333	0.277
19	0.308	0.150	0.542	Y19	0.390	0.333	0.277
20	0.373	0.105	0.522	Y20	0.818	0.032	0.149
21	0.275	0.125	0.600	Y21	0.165	0.169	0.667
22	0.325	0.125	0.550	Y22	0.500	0.167	0.333
23	0.325	0.125	0.550	Y23	0.500	0.167	0.333
24	0.325	0.100	0.575	Y24	0.497	0.000	0.503
25	0.250	0.150	0.600	Y25	0.000	0.333	0.667

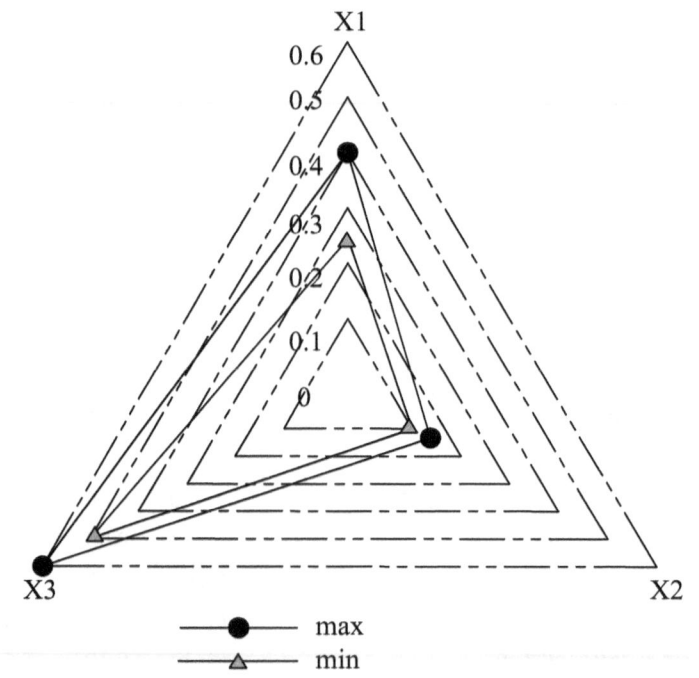

Fig. 6.5.13 Array factor space of the 3- component simplex of soil stabilization

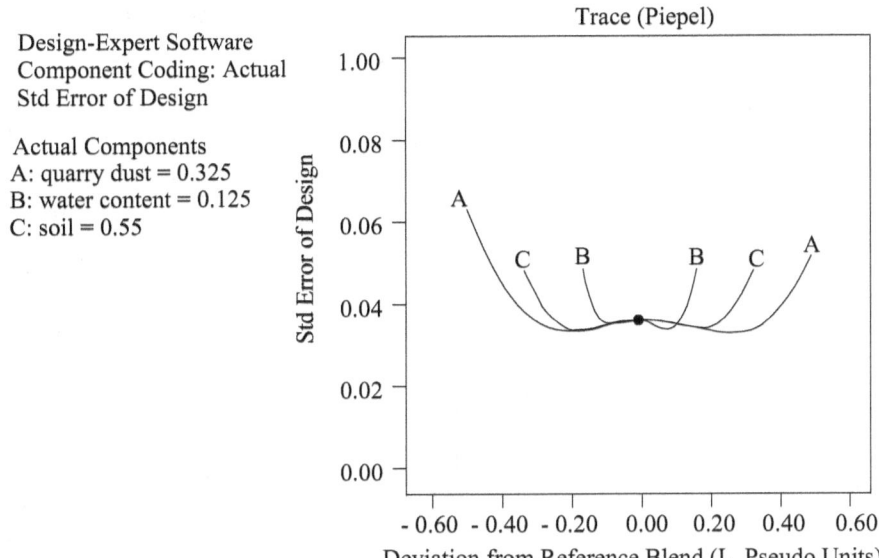

Fig. 6.5.14. Trace and deviation factor space of the 3- component mixture for soil stabilization

Table 6.5.7.16. 2- Component experimental mix proportions for concrete modification

	actual components			pseudo components	
runs	z1	z2	response	x1	x2
1	0.315	0.685	Y1	0.500	0.500
2	0.280	0.720	Y2	0.000	1.000
3	0.333	0.668	Y3	0.750	0.250
4	0.303	0.697	Y4	0.333	0.667
5	0.298	0.703	Y5	0.250	0.750
6	0.327	0.673	Y6	0.667	0.333
7	0.350	0.650	Y7	1.000	0.000

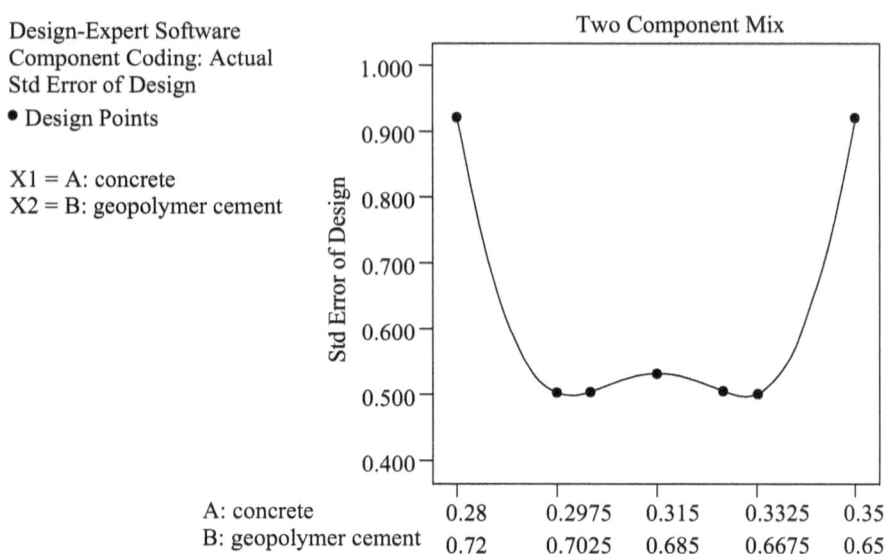

Fig. 6.5.15. Trace and deviation factor space of the 2- component mixture for homogenous mixtures

CHAPTER SEVEN

ENVIRONMENTAL SOILS RE-ENGINEERING

Harnessing the resources of the environment and its hazards is the new age technique of geo-environmental engineering where solid waste materials form a locus that drives ecofriendly and eco-efficient geotechnical engineering practice. We live in an age where every construction activity exposes the earth further to the danger of oxides of carbon emission. Then again professionals are working hard to totally eliminate the utilization of ordinary Portland cement and its attendant carbon dioxide emission dangers. Sustainability in the practice of environmental friendly soils re-engineering is the goal of every professional today and this book has focused its practices on same. Practitioners who will be lucky to pick up this book will be opened to a new horizon of technological advancement in the utilization of supplementary cementitious materials derived from the crushing or direct combustion of solid waste. The environment has to be cared for and there are no better group of persons to do this then the same group causing its depletion. Here we are the technologies and advancement to solving this puzzle. When the geotechnical engineering expert for want of environmental friendly geomaterials rids the environment the indiscriminately dumped solid waste to achieve an eco-efficient and eco-friendly and yet a sustainable soils re-engineering, the earth would be better for it.

CHAPTER EIGHT

CONCLUSION

Generally, the existence of quarry dust in lateritic soils caused the enhancement of unconfined compressive strength and dry density of treated soils, whereas the swelling potentials were found to be significantly decreased with increasing of quarry dust based geopolymer cement (QDbGPC) proportion by weight. The opposite tendency was exhibited when examining drying shrinkage of treated soils, the larger QDbGPC used, the higher value of drying shrinkage limit is seen. Additionally, the consistency limits of three different test soils were reduced with increase in additive proportion. Lastly, the influence of quarry dust coupled material on erodibility potential of treated soils was presented.

Sorptivity and the related conditions of swelling, shrinkage, compression and durability of treated test soils A, B, and C under a hydraulically bound environment have been investigated under the laboratory condition. QD was utilized in proportions by weight to treat the test soils. Increased proportion of QD improved the behaviour of the treated soils in terms of swelling, shrinkage, compression and durability. QD served as additive binder for the soil blend and exhibited pozzolanic properties in the stabilization protocol. Sorptivity behaviour examination has shown to be serious factor to be studied in predicting the service of pavement infrastructures under moisture bounder medium. However, QD has again proved to a good additive in stabilizing soils utilized a pavement foundation materials because its property of withstanding the effect of constant moisture exposure.

Solid waste recycling and reuse as a hub to achieving a more ecofriendly, ecoefficient, and sustainable civil and mechanical engineering infrastructure and a low or zero carbon emission into our planet has been aptly captured in this text. This book among other things has developed a model for effective solid waste combustion, entrapping CO_2 and CO releasing hitherto, and the release of environmental friendly residue of baking soda ($NaHCO_3$), soda ash (Na_2CO_3) and hydrogen gas (H_2). This research has shown the efficiency of using these ecofriendly materials in;

- Soil stabilization to improve the engineering properties of soft clay and expansive soils
- Concrete modification to improve the sulfate, heat and temperature resistance, crack potentials, durability potentials, flexural potentials, compressive strength, etc. of concrete mixes.
- Asphalt modification to improve the frost susceptibility, moisture resistance, heat resistance, sulfate resistance, lateral and axial deformation potentials, and durability potentials of pavement foundations.
- Wastewater treatment to improve on the detoxification process less of CO_2 and volatile gas emissions.

And generally, the sustainability of these operations has been assured and defined by the results of this book.

The analysis of variance (ANOVA) is the tool to be adopted to test for validity and adequacy of the experimental and mathematical modeling operation. With a tested hypothesis under 95% confidence level, the design of experimental protocol would be validated or not. The test for adequacy of the model is usually done using Fischer test at 95% confidence level on the behavioral properties being studied. In this test, two hypotheses would be set as follows:

Null Hypothesis: this states that "there is no significant difference between the laboratory tests and model predicted and the Alternative Hypothesis: states as follows "there is a significant difference between the laboratory test and model predicted". A two-tail test (inequality) will be conducted in this case and if t Stat < -t Critical two-tail or t Stat > t Critical two-tail, we reject the null hypothesis. In ANOVA validation of designs, if F > F crit, we reject the null hypothesis. The developed models can also be tested by writing a representative MATLAB program and observe the running efficiency of the program.

This work has reviewed the use of extreme vertex design in the modeling of the behavior of multicomponent mixture of mixture experiments in civil engineering and composite materials formulation of mechanical engineering designs. Four special cases were cited which were 5-, 4-, 3-, and 2- component mixture experiments of concrete production, asphalt production, soil stabilization and concrete improvement or water treatment exercises. It has shown that these cases can be extrapolated to deal with similar cases in not only civil engineering designs but also in materials engineering, agricultural and bio-resources engineering, chemical engineering, mechanical engineering, polymer and textile engineering, optimization of most production operations in engineering, etc. The cases reviewed yielded results that would eventually guide future users of this optimization technique in civil engineering works and other mixture component modeling works as a hub. The development of constraints is an interesting part of this exercise because it helped in defining the factor space within which experimental points are to be studied for optimal mixture effects.

These mathematical procedures can be adapted engineering problems in Civil, Mechanical, Chemical, and Agricultural Engineering as well as allied disciplines.

REFERENCES

AASHTO (1993). Guide for Design of Pavement Structures. American Association of State Highway and Transportation Officials (AASHTO), Washington DC.

AASHTO (2005). Standard Specification for Transportation Materials and Methods of Sampling and Testing, Part II Methods of Sampling and Testing 25th Edition. American Association of State Highway and Transportation Officials, Washington DC.

AASHTO T 190-09 (2014). Standard method of test for resistance R-value and expansion pressure of compacted soils. American Association of State Highway and Transportation Officials, Washington DC.

AASHTO T 307 (2014). Standard method of test for determining the resilient modulus of soils and aggregate materials. American Association of State Highway and Transportation Officials, Washington DC.

Abdalla, M. S., Jasim, M. A., & Amer, M. I., 2018. Effects of Temperature in Different Initial Duration Time for Soft Clay Stabilized by Fly Ash Based Geopolymer. Civil Engineering Journal, Vol. 4 (9), 2082-2096. https://doi.org/10.28991/cej-03091141

Abdel-Gawwadm, H. A., & Abo-El-Enein, S. A., 2016. A Novel Method to Produce Dry Polymer Cement Powder. *HBRC Journal.* Vol. 12, Pp. 13-24. http://dx.doi.org/10.1016/j.hbrcj.2014.06.0018

Abdul M., Abdullah, Huzaim Sofyan & Irfandi, Safwan, 2015. Utilization of palm oil fuel ash (POFA) in producing lightweight foamed concrete for non-structural building material. The 5th

International Conference of Euro Asia Civil Engineering Forum (EACEF-5), Procedia Engineering 125, 739 – 746. doi: 10.1016/j.proeng.2015.11.119

A. Hawa, Danupon Tonnayopas, & Woraphot Prachasaree, 2013. Performance Evaluation and Microstructure Characterization of Metakaolin-Based Geopolymer Containing Oil Palm Ash. The ScientificWorld Journal, Volume 2013, Article ID 857586, 9 pages. http://dx.doi.org/10.1155/2013/857586

Abubakar, B. H., Putrajaya, R. and Abdulaziz, H., "Malaysian Rice Husk – Improving the Durability and Corrosion Resistance of Concrete: Pre-Review", Vol. 1 (1), 2010, pp. 6-13.

Abood, T.T., Kasa, A.B., & Chik, Z.B. (2007). Stabilization of Silty Clay Soil Using Chloride Compounds, JEST, Malaysia, vol. 2, pp. 102-103

Adrian O. Eberemu, Agapitus A. Amadi & Joseph E. Edeh, (2012). Diffusion of municipal waste contaminants in compacted lateritic soil treated with bagasse ash. Environmental Earth Sciences, DOI 10.1007/s12665-012-2168-z

Ahmad, A. A., Hameed, B. H., & Aziz, N. 2007. Adsorption of direct dye on palm ash: kinetic and equilibrium modeling. Journal of Hazardous Materials, vol. 141 (1), pp. 70-76. https://doi.org/10.1016/j.jhazmat.2006.06.094

Akbari, H., Mensah-Biney, R. & Simms, J., 2015. Production of Geopolymer Binder from Coal Fly Ash to Make Cement-less Concrete. World of Coal Ash (WOCA) Conference in Nasvhille, TN-May 5-7. [online].

A.Ghorbani, & M. Salimzadehshooiili, 2018. Evaluation of Strength behaviour of Cement-RHA Stabilized and Polypropylene Fiber Reinforced Clay-Sand Mixtures. Civil Engineering Journal, Vol. 4 (11), 2628-2641. https://doi.org/10.28991/cej-03091187

Amadi, A. A. & Eberemu, A. O., 2012. Delineation of compaction criteria for acceptable hydraulic conductivity of lateritic soil-bentonite mixtures designed as landfill liners. Environmental Earth Sciences, 67:999–1006. DOI 10.1007/s12665-012-1544-z

American Standard for Testing and Materials (ASTM) C618, 1978. Specification for Pozzolanas. ASTM International, Philadelphia, USA.

A. D'Alessandro, Claudia Fabiani, Anna Laura Pisell, Filippo Ubertini, A. Luigi Materazzi1 & Franco Cotana, 2017. Innovative concretes for low-carbon constructions: a review. International Journal of Low-Carbon Technologies, 12, 289–309. doi:10.1093/ijlct/ctw013

A. O. Eberemu, K. J. Osinubi, T. S. Ijimdiya, & J. E. Sani, 2018. Cement Kiln Dust: Locust Bean Waste Ash Blend Stabilization of Tropical Black Clay for Road Construction. Geotechnical and Geological Engineering, an International Journal, https://doi.org/10.1007/s10706-018-00794-w

A.O. Eberemu, Agapitus A. Amadi & Joseph E. Edeh, 2012. Diffusion of municipal waste contaminants in compacted lateritic soil treated with bagasse ash. Environmental Earth Sciences, DOI 10.1007/s12665-012-2168-z

A.O. Eberemu, A. A. Amadi, & K. J. Osinubi, 2012. The Use of Compacted Tropical Clay Treated With Rice Husk Ash as a Suitable Hydraulic Barrier Material in Waste Containment Applications. Waste and Biomass Valorization, DOI 10.1007/s12649-012-9161-3

A.O. Eberemu. Deborah I. Omajali & Zubairu Abdulhamid, 2015. Effect of Compactive Effort and Curing Period on the Compressibility Characteristics of Tropical Black Clay Treated with Rice Husk Ash. Geotechnical and Geological Engineering, An International Journal, DOI 10.1007/s10706-015-9946-9

A.O. Eberemu, & H. Sada, 2013. Compressibility Characteristics of Compacted Black Cotton Soil Treated With Rice Husk Ash. Nigerian Journal of Technology (NIJOTECH), 32 (3). 507 – 521.

A. O. Eberemu, Joseph.O.Afolayan, Idris. Abubakar, & Kolawole J. Osinubi, 2014. Reliability Evaluation of Compacted Lateritic Soil Treated With Bagasse Ash as Material for Waste Land Fill Barrier. Geo-Congress 2014 Technical Papers, GSP 234, 911-920.

Arioz, O., Tuncan, M., Arioz, E., & Kilinc, K. 2006. Geopolymer: A New Generation Construction Material. 31st Conference on

Our World in Concrete and Structures: 16-17 August, Singapore [online].

Arulrajah, A., Kua, T. A., Horpibulsuk, S., Phetchuay, C., Suksiripattanapong, C., & Du, Y. J. (2016). Strength and microstructure evaluation of recycled glass-fly ash geopolymer as low-carbon masonry units. *Construction and Building Materials*, 114, 400-406.

Ashfaque, A. J., Wan Inn Goh, Kim Hung Mo, Samiullah Sohu, & Imtiaz Ali Bhatti, 2019. Green and Sustainable Concrete – The Potential Utilization of Rice Husk Ash and Egg Shells. Civil Engineering Journal, Vol. 5 (1), 74-81. https://doi.org/10.28991/cej-2019-03091226

Ashfaque A. J., Samiullah Sohu, Muhammad Tahir Lakhiar, Jam Shahzaib, & Ahsan Ali Buriro, 2018. Effectiveness of Locally Available Superplasticizers on the Workability and Strength of Concrete. Civil Engineering Journal, Vol. 4 (12), 2919-2925. https://doi.org/10.28991/cej-03091208

A. G. Lasaki, Reza Jamshidi Chenari, Javad Shamsi Sosahab, & Yaser Jafarian, 2018. Investigation of Strength Parameters of PVA Fiber-Reinforced Fly Ash-Soil Mixtures in Large-Scale Direct Shear Apparatus. Civil Engineering Journal, Vol. 4 (11), 2618-2627. https://doi.org/10.28991/cej-03091186

A.S.M. Abdul Awal & M. Warid Hussin, 2011. Effect of Palm Oil Fuel Ash in Controlling Heat of Hydration of Concrete. The Twelfth East Asia-Pacific Conference on Structural Engineering and Construction, Procedia Engineering 14, 2650–2657. doi:10.1016/j.proeng.2011.07.333

Bui Van, D. & Onyelowe, K.C., 2018. Adsorbed complex and laboratory geotechnics of Quarry Dust (QD) stabilized lateritic soils. Environmental Technology and Innovation, Vol. 10, pp. 355-368. https://doi.org/10.1016/j.eti.2018.04.005.

Bui Van, D., Onyelowe, K.C. & Nguyen Van, M., 2018. Capillary rise, suction (absorption) and the strength development of HBM treated with QD base Geopolymer. International Journal of Pavement

Research and Technology [in press]. https://doi.org/10.1016/j.ijprt.2018.04.003.

Chong, Mei Fong, Lee, Kah Peng, Chieng, Hui Jiun, Syazwani Binti Ramli, & Ili Izyan. 2009. Removal of boron from ceramic industry wastewater by adsorption-flocculation mechanism using palm oil mill boiler (POMB) bottom ash and polymer. Water Research, vol. 43 (13), Pp. 3326-3334. https://doi.org/10.1016/j.watres.2009.04.044

C. N. Ezugwu. 2015. New approaches to solid waste management. Proceedings of the world congress on engineering and computer science, vol. 11, WCECS 2015, Oct. 21-23, 2015, San Francisco, USA.

Chao-Lung, H., Anh-Tuan, B. L., and Chun-Tsun, C. Effect of RHA on the strength and durability characteristics of concrete. Construction and Building Materials, vol. 24 (9), 2011, pp. 3768-3772. https://doi.org/10.1016/j.conbuildmat.2011.04.009

Cornell, John, Experiments with Mixtures: Designs, Models, and the Analysis of Mixture Data, John Wiley & Sons, Inc. New York. 2002.

D. A. Nethercot, 2011. Design of Building Structures to Improve their Resistance to Progressive Collapse. The Twelfth East Asia-Pacific Conference on Structural Engineering and Construction, Procedia Engineering 14, 1–13. doi:10.1016/j.proeng.2011.07.001

D. Mostonejad, S. Noorpour, M. Noorpour, R. Karbati Asl, V. Sadeghi Balkanlouc & A. Karbati Asl. 2017. Effects of petrochemical wastes incinerator ash powder instead of Portland cement on the properties of concrete. Scientia Iranica A, 24(3), 1017-1026

Design Expert 11. Design of experiment software, Stat-Ease Inc., Minneapolis, USA. 2018.

Eberemu, Adrian O., 2015. Compressibility Characteristics of Compacted Lateritic Soil Treated with Bagasse Ash. Jordan Journal of Civil Engineering, 9 (2), 214-228.

Eberemu, A. O., 2013. Evaluation of bagasse ash treated lateritic soil as a potential barrier material in waste containment application. Acta Geotechnica, DOI 10.1007/s11440-012-0204-5

Eberemu, A. O., 2011. Desiccation Induced Shrinkage of Compacted Tropical Clay Treated with Rice Husk Ash. *International Journal of Engineering Research in Africa*, 6, 45-64. doi:10.4028/www.scientific.net/JERA.6.45

Eberemu, A. O., 2011. Consolidation Properties of Compacted Lateritic Soil Treated with Rice Husk Ash. *Geomaterials*, 1, 70-78 doi:10.4236/gm.2011.13011

Eberemu, A. O. & Osinubi, K. J., 2010. Comparative Study of Soil Water Characteristic Curves of Compacted Bagasse Ash Treated Lateritic Soil. 6th International Congress on Environmental Geotechnics, 1378-1383, New Delhi, India

Eberemu, Adrian. O., Edeh, Joseph. E. & Gbolokun, A. O., 2013. The Geotechnical Properties of Lateritic Soil Treated with Crushed Glass Cullet. Advanced Materials Research Vol. 824, pp 21-28. doi:10.4028/www.scientific.net/AMR.824.21

Eberemu, Adrian. O., Amadi, A. A., Garba, Zaharaddeen A., Ibrahim, B., Yusuf, Mohammed, J., Hassan, U. S. & Osinubi, K. J., 2014. Effects of Curing Period and Compactive Effort on the Swelling Pressure of Expansive Soil. *International Conference on Structural and Geotechnical Engineering, Ain Shams University* ICSGE 14.

Eberemu, Adrian. O., Tukka, Diana. D., & Osinubi, Kolawole. J., 2014. The Potential Use of Rice Husk Ash in the Stabilization and Solidification of Lateritic Soil Contaminated with Tannery Effluent. ASCE Geo-Congress 2014 Technical Papers, GSP 234, 2263-2272.

Edeh, J. E., Eberemu, A. O., Abah, A. B., 2012. Reclaimed Asphalt Pavements-Lime Stabilization of Clay as Highway Pavement Materials. *Journal of Sustainable Development and Environmental Protection, 2 (3), Pp. 62-75.*

Edeh, Joseph Ejelikwu, Eberemu, Adrian Oshioname, and Arigi, Abraham S. D., 2012. Reclaimed Asphalt Pavement Stabilized Using Crushed Concrete Waste as Highway Pavement Material, Advances in Civil Engineering Materials, 1 (1), 1–14, DOI: 10.1520/ACEM20120005.

E.R. Sujatha, E. Lakshmipriya, A.R. Sangavi, & K.V., 2018. Poonkuzhali. Inuence of random inclusion of treated sisal fibres on the unconfined compressive strength of highly compressible clay. Scientia Iranica A, 25(5), 2517-2524

Ezugwu, C. N. 2015. Ten Segments of a Comprehensive and Cost-Effective Solid Waste Management System. The Nnewi Engineer. A Quarterly Newsletter of the Nig. Soci. of Engineers, Nnewi Branch, pp. 14-16, June 2015 Edition.

Ezugwu, C. N. 2009. Sanitary Landfilling As an Environmentally Friendly and Cost EffectiveMethod of Waste Disposal. Journal of Civil and Environmental SystemsEngineerig, vol. 10, No. 2, pp. 06-19.

Ezugwu, C. N., Uneke L. A. & Akpan, P. P. 2015. Rice Husk Ash-An alternative to Gypsum in POP board. Inter. Journal of Engineering, Science and Mathematics, vol. 4, Issue 4, pp. 24-35

Fedrigo W., Nunez, W. P., Kleinert T. R., Matuella, M. F. & Ceratti, J. A. P., 2017. Strength, Shrinkage, Erodibility and Capillary Flow Characteritics of Cement-treated Recycled Pavement Materials. International Journal of Pavement Research and Technology, 10, 393-402. http://dx.doi.org/10.1016/j.ijprt.2017.06.001

Foo, K. Y., & Hameed, B. H., 2009. Value-added utilization of oil palm ash: a superior recycling of the industrial agricultural waste. Journal of Hazardous Materials, 172 (2-3), Pp. 523-531. https://doi.org/10.1016/j.jhazmat.2009.07.091

G. Annadural, Ruey-Shin Juang, & D. J. Lee, 2003. Adsorption of heavy metals from water using banana and orange peels. Water Science and Technology, 47 (1), 185-190. DOI: 10.2166/wst.2003.0049

G. Dhinakaran & B. Sreekanth. 2018. Physical, mechanical, and durability properties of ternary blend concrete. Scientia Iranica A, 25(5), 2440-2450

Gartner, E. Industrial interesting approaches to low-CO2 cements. Cement and Concrete Research, 34, 2004, pp. 1489–98.

Gidigasu, M. D., & Dogbey, J. L. K., 1980., Geotechnical Characterization of Laterized Decomposed Rocks for Pavement Construction in Dry Sub-humid Environment". 6th South East

Asian Conference on Soil Engineering, Taipei, 1, 493-506.Garg, S.K. (2005). Soil Mechanics and Foundation Engineering, 6th Edition, Kharna Publishers, Delhi

Goltermann, P., Johansen, V. and Palbol, L., "Packing of aggregates: an alternative tool to determine the optimal aggregate mix", ACI Mat. J. 94(5), 1997, pp. 435-443.

H. Siswanto, B. Supriyanto, Pranoto, and P. R. Chandra, A. R. Hakim. Marshall Properties of asphalt concrete using crumb rubber modified of motorcycle tire waste. *Green Construction and Engineering Education for Sustainable Future,* AIP Conf. Proc. 1887, 020039-1–020039-5; 2017. doi: 10.1063/1.5003522

Hamidi, R. M., Man, Z., & Azizli, K. A., 2016. Concentration of NaOH and the Effect on the Properties of Fly Ash Based Geopolymer. 4th International Conference of Process Engineering and Advanced Materials; Procedia Engineering, Vol. 148, Pp. 189-193. http://dx.doi.org/10.1016/j.proeng.2016.06.568

Hariz, Z., Mohd-MustafaAl-Bakri, A., Kamarudin, H., Nurliyana, A., & Ridho, B. 2017. Review of Various Types of Geopolymer Materials with the Environmental Impact Assessment. MATEC Web of Conferences, Vol. 97, 01021. http://dx.doi.org/10.1051/matecconf/20179701021

Ho, Wilson Wei Sheng, Ng, Hoon Kiat, Gan, Suyin, 2012. Development and characterisation of novel heterogeneous palm oil mill boiler ash-based catalysts for biodiesel production. Bioresources Technology, Vol. 125, pp. 158-164. https://doi.org/10.1016/j.biortech.2012.08.099

Hoy, M., Horpibulsuk, S., Rachan, R., Chinkulkijniwat, A., & Arulrajah, A., 2016. Recycled asphalt pavement–fly ash geopolymers as a sustainable pavement base material: Strength and toxic leaching investigations. *Science of the Total Environment*, 573, 19-26.

Ibrahim H., Wael Seddik & Hossam Ibrahim, 2017. Towards green cities in developing countries: Egyptian new cities as a case study. International Journal of Low-Carbon Technologies, 12, 358–368. doi:10.1093/ijlct/ctx009

Ikeagwuani, C. C. & Nwonu, D. C., 2019. Emerging trends in expansive soil stabilization; a review. Journal of Rock Mechanics and Geotechnical Engineering. https://doi.org/10.1016/j.jrmge.2018.08.013

Ikeagwuani, C. C. & Obeta, I. N., 2019. Stabilization of black cotton soil subgrade using sawdust ash and lime. Soils and Foundations (in press). https://doi.org/10.1016/j.sandf.2018.10.004

I. S. Muhammad, A. O. Eberemu, A. M. Kundiri & A. S. Muhammed, 2018. Remediation Techniques for contaminated Soils: A Review. University of Maiduguri Faculty of Engineering Seminar Series, 9 (2), November 2018, Pp. 52-61.

J. E. Sani, P. Yohanna, K. R. Etim, J. K. Osinubi, & O. A. Eberemu, 2017. Reliability Evaluation of Optimum Moisture Content of Tropical Black Clay Treated with Locust Bean Waste Ash as Road Pavement Sub-base Material. Geotechnical and Geological Engineering, an International Journal, 35 (5), 35:2421-2431. DOI 10.1007/s10706-017-0256-2

J. A. Herdt, John Hunt & Kellen Schauermann, 2016. Newly invented biobased materials from low-carbon, diverted waste fibers: research methods, testing, and full-scale application in a case study structure. International Journal of Low-Carbon Technologies, 11, 400–415. doi:10.1093/ijlct/ctw018

J. R. Oluremi, A. O. Eberemu, T. S. Ijimdiya, & K. J. Osinubi, 2016. Absorption and Diffusion Potential of Waste Wood Ash-Treated Lateritic Soil. ASCE Geo-Chicago 2016 GSP 273, 98-106.

J. R. Oluremi, Paul Yohanna, Kazeem Ishola, Godwin Lazhi Yisa, Adrian O. Eberemu, Stephen T. Ijimdiya & Kolawole J. Osinubi, 2017. Plasticity of Nigerian lateritic soil admixed with selected admixtures. Environmental Geotechnics http://dx.doi.org/10.1680/jenge.15.00085, Paper 15.00085

J. R. Oluremi, Stephen T. Ijimdiya, Adrian O. Eberemu & Kolawole J. Osinubi, 2018. Reliability Evaluation of Hydraulic Conductivity Characteristics of Waste Wood Ash Treated Lateritic Soil. Geotechnical and Geological Engineering, an International Journal, https://doi.org/10.1007/s10706-018-0625-5

Kamal Rahmani, Mohsen Ghaemian & Seyed Abbas Hosseini, 2019. Experimental study of the effect of water to cement ratio on mechanical and durability properties of Nano-silica concretes with Polypropylene fibers. Scientia Iranica A, (online) DOI: 10.24200/sci.2017.5077.1079

Kennedy Chibuzor Onyelowe, Duc Bui Van, Manh van Nguyen, & Henry Ugwuanyi. 2018. Effect of Ceramic Waste Derivatives on the Volume Change Behavior of Soft Soils For Moisture Bound Transport Geotechnics. Electronic Journal of Geotechnical Engineering, (23.04), pp 821-834

Khairunisa Muthusamya, Nurazzimah Zamria, Mohammad Amirulkhairi Zubira, Andri Kusbiantorob, & Saffuan Wan Ahmad, 2015. Effect of mixing ingredient on compressive strength of oil palm shell lightweight aggregate concrete containing palm oil fuel ash. The 5th International Conference of Euro Asia Civil Engineering Forum (EACEF-5), Procedia Engineering 125, 804 – 810. doi: 10.1016/j.proeng.2015.11.142.

Kim Hung Moa, U. Johnson Alengarama, & Mohd Zamin Jumaat, 2015. Experimental investigation on the properties of lightweight concretecontaining waste oil palm shell aggregate. The 5th International Conference of Euro Asia Civil Engineering Forum (EACEF-5), Procedia Engineering 125, 587 – 593. doi: 10.1016/j.proeng.2015.11.065

K. C. Onyelowe, Duc Bui Van, Francis Orji, Manh Nguyen Van1b, Clifford Igboayaka and Henry Ugwuanyi. "Exploring Rock by Blasting with Gunpowder as Explosive, Aggregate Production and Quarry Dust Utilization for Construction Purposes" Oklahoma State University *Electronic Journal of Geotechnical Engineering*, (23.04), 2018, pp 447-456. http://www.ejge.com/2018/Ppr2018.0112ma.pdf

K. C. Onyelowe. Review on the role of solid waste materials in soft soils reengineering. Materials Science for Energy Technologies, Vol. 2 (1), 2019, pp. 46-51. https://doi.org/10.1016/j.mset.2018.10.004

K. C. Onyelowe, G. Alaneme, C. Igboayaka, F. Orji, H. Ugwuanyi, D. Bui Van, M. Nguyen Van, Scheffe optimization of swelling,

California bearing ratio, compressive strength, and durability potentials of quarry dust stabilized soft clay soil, Materials Science for Energy Technologies, vol. 2(1), 2019, pp. 67-77. https://doi.org/10.1016/j.mset. 2018.10.005

K. J. Osinubi, Adrian O. Eberemu, T. Stephen Ijimdiya, John E. Sani, & S. E. Yakubu, 2018. Volumetric Shrinkage of Compacted Lateritic Soil Treated with Bacillus pumilus. Proceedings of GeoShanghai 2018 International Conference: Geoenvironment and Geohazard, 315–324, 2018. https://doi.org/10.1007/978-981-13-0128-5_36

K. J. Osinubi, A. O. Eberemu, P. Yohanna & R. K. Etim, 2016. Reliability Estimate of the Compaction Characteristics of Iron Ore Tailings Treated Tropical Black Clay as Road Pavement Sub-Base Material. ASCE Geo-Chicago 2016 GSP 271, 855-864.

K. J. Osinubi, Johnson R. Oluremi, Adrian O. Eberemu & Stephen T. Ijimdiya, 2017. Interaction of landfill leachate with compacted lateritic soil–waste wood ash mixture. Proceedings of the Institution of Civil Engineers, Waste and Resource Management, http://dx.doi.org/10.1680/jwarm.17.00012 Paper 1700012

K.J. Osinubi, R. Diden, T.S. Ijimdiya and A.O. Eberemu, 2012. Effect of Elapsed Time after Mixing on the Strength Properties of Black Cotton Soil Cement Blast Furnace Slag Mixtures. Journal of Engineering Research, 17 (4), 63-75.

K.J. Osinubi, P. Yohanna, A.O. Eberemu, 2015. Cement modification of tropical black clay using iron ore tailings as admixture. Transportation Geotechnics, 5, 35-49. http://dx.doi.org/10.1016/j.trgeo.2015.10.001

K. J. Osinubi, Adrian O. Eberemu, & Olusegun B. Akinmade, 2016. Evaluation of Strength Characteristics of Tropical Black Clay Treated with Locust Bean Waste Ash, Geotechnical and Geological Engineering. DOI 10.1007/s10706-015-9972-7

K.J. Osinubi, A.O. Eberemu, A.O. Bello, & A. Adzegah, 2012. Effect of Fines Content on the Engineering Properties of Reconstituted Lateritic Soils in Waste Containment Application. Nigerian Journal of Technology (NIJOTECH), 31 (3), pp. 277-287.

K. J. Osinubi & A. O. Eberemu, 2013. Hydraulic conductivity of compacted lateritic soil treated with bagasse ash. International Journal of Environment and Waste Management, 11 (1), 38-58.

K. J. Osinubi, Adrian O. Eberemu & Agapitus A. Amadi, 2012. Compatibility of compacted lateritic soil treated with bagasse ash and Municipal Solid Waste leachate. International Journal of Environment and Waste Management, 10 (4), 365-376.M. A. Barakat, 2011. New trends in removing heavy metals from industrial wastewater. Arabian Journal of Chemistry, 4, 361–377. doi:10.1016/j.arabjc.2010.07.019

M. Zhang, and V. M. Malhotra, "High-Performance Concrete Incorporating Rice Husk Ash as a Supplementary Cementing Material", ACI Materials Journal, November-December 1996, Tittle no. 93-M72, pp. 629-636.

Mahmud, M.I. and Cho, H.M., 2018. A review on characteristics, advantages and limitations of palm oil biofuel, *Int. J. Global Warming*, 14 (1), 81–96.

Maryoto, Nor Intang Setyo Hermanto, Yanuar Haryanto, Sugeng Waluyoa, & Nur Alvi Anisa, 2015. Influence of prestressed force in the waste tire reinforced concrete. The 5th International Conference of Euro Asia Civil Engineering Forum (EACEF-5), Procedia Engineering 125, 638 – 643. doi: 10.1016/j.proeng.2015.11.088

M. Ehsani, N. Shariatmadari & S.M. Mirhosseini. 2017. Experimental study on behavior of soil-waste tire mixtures. Scientia Iranica A, 24(1), 65-71

M.K. Atahu, F. Saathoff, & A. Gebissa, 2019. Mechanical behaviors of expansive soil treated with coffee husk ash. Journal of Rock Mechanics and Geotechnical Engineering. https://doi.org/10.1016/j.jrmge.2018.11.004

Mehta, P.K., "High-performance, high-volume fly ash concrete for sustainable development", Proceedings of the International Workshop on Sustainable Development and Concrete Technology, Beijing, China, May 20–21, 2004.

Minitab 18. Minitab statistical software, Minitab Inc., Pennsylvania, USA. 2018.

Mohammad, H. R., Milad Sadeghzadeh, Mohammad Alhuyi Nazari, Mohammad Hossein Ahmadi & Fatemeh Razi Astaraei, 2018. Applying GMDH artificial neural network in modeling CO2 emissions in four Nordic countries. International Journal of Low-Carbon Technologies, 13, 266–271. doi:10.1093/ijlct/cty026

Mohammadreza, A., Amir Izadi, Fereshte Asadiamiri, 2018. Investigating Effects of Coal Flotation Waste on Aged Hot Mix Asphalt Performance. Civil Engineering Journal, Vol. 4 (10), 2491-2501. https://doi.org/10.28991/cej-03091175

M. Roustaei, M. Ghazavi & E. Aliaghaei, 2016. Application of tire crumbs on mechanical properties of a clayey soil subjected to freeze-thaw cycles. Scientia Iranica A, 23(1), 122-132

Muhammad K. & Hamid N., 2015. Using advanced materials of granular BRA modifier binder to improve the flexural fatigue performance of asphalt mixtures. The 5th International Conference of Euro Asia Civil Engineering Forum (EACEF-5), Procedia Engineering 125, 452 – 460. doi: 10.1016/j.proeng.2015.11.120

Muhammad S. K., Muhammad Tufail, Mateeullah Mateeullah, 2018. Effects of Waste Glass Powder on the Geotechnical Properties of Loose Subsoils. Civil Engineering Journal, Vol. 4 (9), 2044-2051. https://doi.org/10.28991/cej-03091137

Naraindas B., Shanker Lal Meghwar, Samiullah Sohu, Ali Raza Khoso, Ashok Kumar, Zubair Hussain Shaikh, 2018. Experimental Study on Recycled Concrete Aggregates with Rice Husk Ash as Partial Cement Replacement. Civil Engineering Journal, Vol. 4 (10), 2305-2314. https://doi.org/10.28991/cej-03091160

Naraindas B., Shanker Lal Meghwar, Suhail Ahmed Abbasi, Lal Chand Marwari, Jabbar Ahmed Mugeri, Rameez Ali Abbasi, 2018. Effect of Rice Husk Ash and Water-Cement Ratio on Strength of Concrete. Civil Engineering Journal, Vol. 4 (10), 2373-2382. https://doi.org/10.28991/cej-03091166

N. Hormaza'bal, Mark Gillott & Brian Ford, 2016. The performance and in-use experience of low to zero carbon technologies in an experimental energy home. International Journal of Low-Carbon Technologies, 11, 283–295. doi:10.1093/ijlct/ctv006

N. Kaid, M. Cyr, & H. Khelafi, 2015. Characterization of an Algerian natural pozzolan for its use in eco-efficient cement. *International Journal of Civil Engineering, Vol. 13, No. 4A, Transaction A: Civil Engineering.* DOI: 10.22068/IJCE.13.4.444

Nesibe, G. O., Bappy Ahsan, Said Mansour, & Srinath R. Iyengar, 2013. Mechanical performance and durability of treated palm fiber reinforced mortars. Gulf Organisation for Research and Development International Journal of Sustainable Built Environment, 2, 131–142. http://dx.doi.org/10.1016/j.ijsbe.2014.04.002.

Normunds, S., Merlin Liiv, Ilze Ozola and Triin Reitalu, 2018. Carbon accumulation rate in a raised bog in Latvia, NE Europe, in relation to climate warming. Estonian Journal of Earth Sciences, 2018, **67**, 4, 247–258 https://doi.org/10.3176/earth.2018.20

O. K. Wattimena, Antoni and D. Hardjito. A Review on the Effect of Fly Ash Characteristics and Their Variations on the Synthesis of Fly Ash Based Geopolymer. *Green Construction and Engineering Education for Sustainable Future,* AIP Conf. Proc. 1887, 020041-1–020041-12; 2017. doi: 10.1063/1.5003524

Onuh, E.I. and Inambao, F.L., 2018 An evaluation of neat biodiesel/diesel performance, emission pattern of NOx and CO in compression ignition engine', *Int. J. Global Warming,* 14 (1), 21–39.

Onyelowe, K. C. & Ubachukwu, O. A., 2015. Stabilization of Olokoro-Umuahia Lateritic Soil using Palm Bunch Ash (PBA) as Admixture, *Umudike Journal of Engineering and Technology* (UJET), Volume 1, Number 2, Pp. 67-77.

Onyelowe, K. C. & Okafor, F. O., 2015. Review of the Synthesis of Nano-Sized Ash from Local Waste for Use as Admixture or Filler in Engineering Soil Stabilization and Concrete Production, *Journal of Environmental Nanotechnology,* (JENT), Vol. 4, Issue 4, Pp. 23-27.

Onyelowe, K. C., 2017b. Nanosized palm bunch ash (NPBA) stabilisation of lateritic soil for construction purposes, *International Journal of Geotechnical Engineering,* TandF online, http://dx.doi.org/10.1080/19386362.2017.1322797

Onyelowe, K. C. 2017c. Nanostructured Waste Paper Ash Stabilization of Lateritic Soils for Pavement Base Construction Purposes.

Electronic Journal of Geotechnical Engineering, (22.09), pp 3633-3647. www.ejge.com

Onyelowe, K.C., 2017d. Solid Wastes Management (SWM) in Nigeria and their Utilization in the Environmental Geotechnics as an Entrepreneurial Service Innovation (ESI) for Sustainable Development. Int J Waste Resour 7: 282. ISSN: 2252-5211.

Onyelowe, K. C. & Maduabuchi M. N., 2017a. Palm Bunch Management and Disposal as Solid Waste and the Stabilization of Olokoro Lateritic Soil for Road Construction Purposes in Abia State, Nigeria. Int J Waste Resour Vol 7 Issue 2. https://doi.org/10.4172/2252-5211.1000279

Onyelowe, K. C., Ekwe, N. P., Okafor, F. O., Onuoha, I. C., Maduabuchi, M. N., & Eze, G. T., 2017c. Investigation of the Stabilization Potentials of Nanosized-Waste Tyre Ash (NWTA) as Admixture with Lateritic Soil in Nigeria. *Umudike Journal of Engineering and Technology (UJET), Vol. 3, No. 1, Pp. 26 – 35.* www.ujetmouau.com

Onyelowe, K. C. & Bui Van, D., 2018a. Durability of nanostructured biomasses ash (NBA) stabilized expansive soils for pavement foundation, International Journal of Geotechnical Engineering, https://doi.org/10.1080/19386362.2017.1422909

Onyelowe, K. C., Duc Bui Van & Manh Nguyen Van, 2018b. Swelling potential, shrinkage and durability of cemented and uncemented lateritic soils treated with CWC base geopolymer, International Journal of Geotechnical Engineering. https://doi.org/10.1080/19386362.2018.1462606

Onyelowe KC & Bui Van D., 2018. Predicting Subgrade Stiffness of Nanostructured Palm Bunch Ash Stabilized Lateritic Soil for Transport Geotechnics Purposes. Journal of GeoEngineering of Taiwan Geotechnical Society, vol. 13(2), pp. 59-67. http://140.118.105.174/jge/article.php?v=20&i=72&volume=13&issue=2

Osinubi, K. J. & Eberemu, A. O., 2019. Compatibility and Attenuative Properties of Blast Furnace Slag Treated Laterite. Journal of Solid Waste Technology and Management, 35(1 8), P*p. 7-16.*

Osinubi, K. J. & Eberemu, A. O., 2010. Unsaturated Hydraulic Conductivity Of Compacted Lateritic Soil Treated With Bagasse Ash. ASCE GeoFlorida 2010: Advances in Analysis, Modeling & Design (GSP 199), 357-369.

Osinubi, K. J. & Eberemu, A. O., 2008. Effect of Desiccation on Compacted Lateritic Soil Treated with Bagasse Ash. Materials Society of Nigeria (MSN) Zaria Chapter Book of Proceedings, 4th Edition. Pp. 1-10.

Osinubi, K. J., Oluremi, J. R., Eberemu, A. O. & Ijimdiya, S. T., 2017. Interaction of landfill leachate with compacted lateritic soil–waste wood ash mixture. Proceedings of the Institution of Civil Engineers, Waste and Resource Management 170, Issue WR3+4, 128–138 https://doi.org/10.1680/jwarm.17.00012

Phetchuay, C., Horpibulsuk, S., Arulrajah, A., Suksiripattanapong, C., & Udomchai, A., 2016. Strength development in soft marine clay stabilized by fly ash and calcium carbide residue based geopolymer. *Applied Clay Science*, 127, 134-142.

Phetchuay, C., Horpibulsuk, S., Suksiripattanapong, C., Chinkulkijniwat, A., Arulrajah, A., & Disfani, M. M., 2014. Calcium carbide residue: Alkaline activator for clay–fly ash geopolymer. *Construction and Building Materials*, 69, 285-294.

P. Kathirvel & S.R.M. Kaliyaperumal, 2018. Performance of alkali activated slag concrete under aggressive environment. Scientia Iranica A, 25(5), 2451-2460

R. A. McLean and V. L. Anderson. Extreme Vertices Design of Mixture Experiments. *Technometrics*, vol. 8, No. 3, 1966, pp. 447-454. https://doi.org/10.1080/00401706.1966.10490377

Rahul, B. & Baboo R., 2019. Efficiency Concepts and Models that Evaluates the Strength of Concretes Containing Different Supplementary Cementitious Materials. Civil Engineering Journal, Vol. 5 (1), 18-32. http://dx.doi.org/10.28991/cej-2019-03091222

Ramakrishna, M., Li Xuanjun, Avner Adin & Suresh Valiyaveettil, 2018. Fruit Peels as Efficient Renewable Adsorbents for Removal of Dissolved Heavy Metals and Dyes from Water. ACS Sustainable

Chemistry & Engineering, Vol. 3 (6), pp. 1117-1124. https://doi.org/10.1021/acssuschemeng.5b00207

Rajak, M. A. A., Majid, Zaiton Abdul, & Ismail, 2015. Mohammad. Physicochemical characterizations of nano-palm oil fuel ash. AIP Conference Proceedings **1669**, 020018; https://doi.org/10.1063/1.4919156

R.K. Etim, A.O. Eberemu, & K.J. Osinubi, 2017. Stabilization of black cotton soil with lime and iron ore tailings admixture. Transportation Geotechnics, 10, 85-95. http://dx.doi.org/10.1016/j.trgeo.2017.01.002

Saad, T., Asad Ullah, Kamal Shah, Faial Mehmood, Akhtar Gul, 2018. Influence of Reduced Water Cement Ratio on Behaviour of Concrete Having Plastic Aggregate. Civil Engineering Journal, Vol. 4 (12), 2971-2977. https://doi.org/10.28991/cej-03091213

Salahudeen, A. B., Eberemu, A. O. & Osinubi, K. J., 2014. Assessment of Cement Kiln Dust-Treated Expansive Soil for the Construction of Flexible Pavements. Geotechnical and Geological Engineering, an International Journal, 32:923-931. DOI 10.1007/s10706-014-9769-0

Scheffé, H. Experiments with mixtures. Journal of the Royal Statistical Society Series B. 20, (1958), 344-360.

Smith, W. F. Experimental design for formulation. American Statistical Association and Society for Industrial and Applied Mathematics, USA. 1931.

Snee, R. D. Experimental designs for mixture systems with multicomponent constraints. Communications in Statistics-Theory and Methods, 8(4), 1979, pp. 303-326. http://dx.doi.org/10.1080/03610927908827762

Soosan, T. G., Sridharan, A., Jose, B. T., & Abraham, B. M., 2005. Utilization of quarry dust to improve the Geotechnical properties of soils in highway construction. ASTM Geotechnical Testing Journal, Vol. 28 (3), pp. 1-10.

Sridharan, A. & Keshavamurthy, P., 2016. Expansive soil characterization; an appraisal. INAE Letters, 1(1), 29-33.

Srinivasan, K., & Sivakumar, A., 2013. Geopolymer Binders: A Need for Future Concrete Construction. *ISRN Polymer Sciences*, Vol. 2013. http://dx.doi.org/10.1155/2013/509185

S. Kumar, Vadym Drozd & Surendra K. Saxena, 2012. Catalytic Studies of Sodium Hydroxide and Carbon Monoxide Reaction. *Catalysts* 2012, 2, 532-543; doi:10.3390/catal2040532

Suksiripattanapong, C., Srijumpa, T., Horpibulsuk, S., Sukmak, P., & Arulrajah, A., 2015. Compressive strengths of water treatment sludge-fly ash geopolymer at various compression energies. *Lowland Technology International*, 17(3), 147-156.

Sukmak, P., Horpibulsuk, S., Shen, S. L., Chindaprasirt, P., & Suksiripattanapong, C., 2013. Factors influencing strength development in clay–fly ash geopolymer. *Construction and Building Materials*, 47, 1125-1136.

T. S. Amhadi and G. J. Assaf. Overview of Soil Stabilization Methods in Road Construction. GeoMEast 2018, SUCI, pp. 21–33, 2019. https://doi.org/10.1007/978-3-030-01911-2_3

T. S. Amhadi and G. J. Assaf. The Effect of Using Desert Sands and Cement to Stabilize the Base Course Layer of Roads in Libya. New Prospects in Geotechnical Engineering Aspects of Civil Infrastructures, Sustainable Civil Infrastructures, 2019, https://doi.org/10.1007/978-3-319-95771-5_12

T. W. Samadhi, W. Wulandari, M. I. Prasetyo, M. R. Fernando, and A. Purbasari. Synthesis of Geopolymer from Biomass-Coal Ash Blends. *Green Construction and Engineering Education for Sustainable Future*, AIP Conf. Proc. 1887, 020031-1–020031-7; 2017. doi: 10.1063/1.5003514

Taha, M., Mohammad Ismail, Salihuddin Radin Sumadi, Muhammad Aamer Rafique Bhutta, Mostafa Samadi, & Seyed Mahdi Sajjadi, 2014. Binary Effect of Fly Ash and Palm Oil Fuel Ash on Heat of Hydration Aerated Concrete. The Scientific World Journal, Volume 2014, Article ID 461241, 6 pages. http://dx.doi.org/10.1155/2014/461241

Taherkhani, H., 2016. Investigating the Properties of Asphalt Concrete Containing Glass Fibers and Nanoclay. *Civil Engineering*

Infrastructures Journal, 49(1): 45 – 58, June 2016. DOI: 10.7508/ceij.2016.01.004

Tangchirapat, W., Saeting Tirasit, Jaturapitakkul Chai, Kiattikomol Kraiwood, & Siripanichgorn Anek, 2007. Use of waste ash from palm oil industry in concrete. Waste Management, 27 (1), Pp. 81-88 https://doi.org/10.1016/j.wasman.2005.12.014

Vikas, S., Atul, Ashhad Imam, P. K. Mehta, Satyendranath, & M. K. Tripathi, 2018. Supplementary Cementitious Materials in Construction - An Attempt to Reduce CO_2 Emission. *J. Environ. Nanotechnol. Vol. 7(2) pp. 31-36. doi:10.13074/jent.2018.06.182306*

W. Wangkananon, C. Phuaksaman, T. Koobkokkruad and S. Natakankitkul. An Extreme Vertices Mixture Design Approach to Optimization of Tyrosinase Inhibition Effects. Engineering Journal, vol. 22 (1), 2018. DOI:10.4186/ej.2018.22.1.175

Wunchock, K., Theerawat Sinsiri, & Chai Jaturapitakkul, 2011. Effect of Palm Oil Fuel Ash Fineness on Packing Effect and Pozzolanic Reaction of Blended Cement Paste. The Twelfth East Asia-Pacific Conference on Structural Engineering and Construction, Procedia Engineering 14, 361–369. doi:10.1016/j.proeng.2011.07.045

Xiaoyan, H., Ravi Ranade, & Victor C. Li, 2013. Feasibility Study of Developing Green ECC Using Iron Ore Tailings Powder as Cement Replacement. Journal of Materials in Civil Engineering, 25 (7), 923-931. DOI: 10.1061/(ASCE)MT.1943-5533.0000674

Yin, C. Y., Wan Ali, Wan S., Lim, & Ying, P., 2008. Oil palm ash as partial replacement of cement for solidification/stabilization of nickel hydroxide sludge. Journal of Hazardous Materials, vol. 150 (2), Pp. 413-418. https://doi.org/10.1016/j.jhazmat.2007.04.119

Zareei, S. A., Ameri, F., Dorostkar, F. and Ahmadi, M. RHA as a partial replacement for cement in high strength concrete containing micro silica: Evaluating durability and mechanical properties. Case Studies in Construction Materials, vol. 7, 2017, pp. 73-81. https://doi.org/10.1016/j.cscm.2017.05.001

APPENDIXES

Apendix 1 Compaction behaviour of treated soil

Proportion of Geopolymer to Ordinary Portland Cements (GPC/OPC)	Percentage of Crushed Watse Glassess (CWG) Added (%)															
	0%			4%			8%			12%			16%			20%		
	MDD	OMC	Gs	MDD	OMC	Gs	MDD	OMC	Gs	MDD	OMC	Gs	MDD	OMC	Gs	MDD	OMC	Gs
0:0	1.85	16.2	2.43	1.87	16.1	2.45	1.89	15.9	2.47	1.92	14.9	2.49	1.98	14.2	2.56	1.99	14.1	2.58
0:40	3.5	15.4	3.45	3.56	15.2	3.48	3.58	15.1	3.49	3.59	13.1	3.54	3.66	12.8	3.58	3.86	12.4	3.68
5:35	4.6	14.3	4.5	4.68	14.1	4.59	4.69	14	4.6	4.72	12.0	4.69	4.89	11.7	4.89	4.99	11.2	4.99
10:30	5.5	13.5	5.34	5.59	13.2	5.37	5.6	13.1	5.38	5.69	11.1	5.88	5.88	10.6	5.98	5.98	10.2	6.1
15:25	6.4	12.6	6.45	6.48	12.3	6.48	6.49	12.2	6.49	6.53	10.2	6.89	6.76	9.4	6.99	6.86	9.1	7.2
20:20	7.5	11.7	7.45	7.58	11.3	7.49	7.59	11.2	7.52	7.64	9.2	8.58	7.87	8.2	8.88	8.87	7.2	9.48
25:15	9.56	9.45	10.4	9.59	9.25	10.5	9.69	7.25	11.5	9.78	6.25	12.58	10.48	6.0	12.88	11.48	5.6	13.78
30:10	11.45	7.34	13.2	11.55	7.14	13.8	11.85	6.14	14.8	11.98	5.14	15.8	12.98	5.0	15.9	13.98	4.2	16.86
35:5	13.65	5.45	15.6	13.85	5.15	15.9	13.95	4.15	16.9	14.24	3.15	17.9	15.24	3.0	18.78	16.24	2.8	19.88
40:0	15.76	4.75	18.5	15.96	4.35	18.8	16.06	3.35	19.8	17.86	2.35	20.8	18.56	2.0	21.8	19.56	2.1	22.64

Appendix 2 Consistency behaviour of treated soil

Proportion of Geopolymer to Ordinary Portland Cements (GPC/OPC)	Percentage of Crushed Watse Glassess (CWG) Added (%)																	
	0%			4%			8%			12%			16%			20%		
	LL	PL	Ip	LL	PL	Ip	LL	PL	Ip	LL	PL	Ip	LL	PL	Ip	LL	PL	Ip
0:0	46	21	25	44	22	24	40	17	23	38	16	22	37	16	21	36	16	20
0:40	43	21	22	42	21	21	38	18	20	37	17	20	35	16	19	34	15	19
5:35	41	20	21	39	19	20	37	18	19	35	17	18	32	15	17	30	13	17
10:30	38	18	20	36	17	19	34	16	18	32	15	17	30	14	16	28	13	15
15:25	35	17	18	32	14	18	29	12	17	26	10	16	25	10	15	24	10	14
20:20	31	15	16	30	15	15	27	13	14	24	11	13	23	11	12	21	10	11
25:15	29	16	13	26	14	12	23	11	11	21	11	10	20	11	9	18	10	8
30:10	25	15	10	22	13	9	19	11	8	17	10	7	15	9	6	14	9	5
35:5	19	12	7	17	11	6	16	11	5	15	11	4	14	11	3	12	10	2
40:0	16	12	4	14	11	3	12	9	3	10	8	2	10	9	1	9	8	1

Appendix 3 California bearing ratio behaviour of OPC+QDbGPC (%) treated soil with 0% CWG

Plunger Penetration (mm)	Plunger Load (kN)								
	California bearing ratio behaviour of OPC+QDbGPC (%) treated soil with 0% CWG								
	0	0+40	5+35	10+30	20+20	25+15	30+10	35+5	40+0
0	0	0	0	0	0	0	0	0	0
0.5	1.2	1.5	1.8	2.1	4.4	8.4	12.4	16.5	20.5
1	1.3	1.6	1.9	2.2	4.5	8.5	12.5	16.6	20.6
1.5	1.5	1.7	2.0	2.3	4.6	8.6	12.6	16.7	20.7
2	1.7	1.8	2.1	2.4	4.7	8.7	12.7	16.8	20.8
2.5	1.8	1.9	2.2	2.5	4.8	8.8	12.8	16.9	20.9
3	2.0	2.1	2.3	2.6	4.9	8.9	12.9	17.0	21.0
3.5	2.1	2.2	2.4	2.7	5.0	9.0	13.0	17.1	21.1
4	2.2	2.3	2.5	2.8	5.1	9.1	13.1	17.2	21.2
4.5	2.3	2.4	2.6	2.9	5.2	9.2	13.2	17.3	21.3
5	2.4	2.5	2.7	3.0	5.3	9.3	13.3	17.4	21.4
5.5	2.6	2.7	2.8	3.1	5.4	9.4	13.4	17.5	21.5
6	2.7	2.8	2.9	3.2	5.5	9.5	13.5	17.6	21.6
6.5	2.9	2.9	3.0	3.3	5.6	9.6	13.6	17.7	21.7
7	3.2	3.3	3.4	3.5	5.7	9.7	13.7	17.8	21.8
7.5	3.6	3.7	3.8	3.7	5.8	9.8	13.8	17.9	21.9
8	3.8	3.9	4.0	4.1	6.2	9.9	13.9	18.0	22.0
8.5	4.1	4.2	4.3	4.4	6.5	10.0	14.0	18.1	22.1
9	4.3	4.4	4.5	4.6	6.7	10.1	14.1	18.2	22.2
9.5	4.4	4.5	4.6	4.7	6.8	10.2	14.2	18.3	22.3
10	4.5	4.6	4.7	4.8	6.9	10.3	14.3	18.4	22.4

Appendix 4 California bearing ratio behaviour of OPC+QDbGPC (%) treated soil with 4% CWG

Plunger Penetration (mm)	Plunger Load (kN)								
	California bearing ratio behaviour of OPC+QDbGPC (%) treated soil with 4% CWG								
	0	0+40	5+35	10+30	20+20	25+15	30+10	35+5	40+0
0	0	0	0	0	0	0	0	0	0
0.5	1.3	1.6	1.9	2.2	4.5	8.5	12.5	16.6	20.6
1	1.4	1.7	2.0	2.3	4.6	8.6	12.6	16.7	20.7
1.5	1.6	1.8	2.1	2.4	4.7	8.7	12.7	16.8	20.8
2	1.8	1.9	2.2	2.5	4.8	8.8	12.8	16.9	20.9
2.5	1.9	2.0	2.3	2.6	4.9	8.9	12.9	17.0	21.0
3	2.1	2.2	2.4	2.7	5.0	9.0	13.0	17.1	21.1
3.5	2.2	2.3	2.5	2.8	5.1	9.1	13.1	17.2	21.2
4	2.3	2.4	2.6	2.9	5.2	9.2	13.3	17.3	21.3
4.5	2.4	2.5	2.7	3.0	5.3	9.3	13.4	17.4	21.4
5	2.5	2.6	2.8	3.1	5.4	9.4	13.5	17.5	21.5
5.5	2.7	2.7	2.9	3.2	5.5	9.5	13.6	17.6	21.6
6	2.8	2.9	3.0	3.3	5.6	9.6	13.7	17.7	21.7
6.5	3.0	3.1	3.1	3.4	5.7	9.6	13.8	17.8	21.8
7	3.3	3.4	3.5	3.6	5.8	9.8	13.9	17.9	21.9
7.5	3.7	3.8	3.9	3.8	5.9	9.9	14.0	18.0	22.0
8	3.9	4.0	4.1	4.2	6.3	10.0	14.1	18.1	22.1
8.5	4.2	4.3	4.4	4.5	6.6	10.1	14.2	18.2	22.2
9	4.3	4.5	4.6	4.7	6.8	10.2	14.3	18.3	22.3
9.5	4.5	4.6	4.7	4.8	6.9	10.3	14.3	18.4	22.4
10	4.6	4.7	4.8	4.9	7.0	10.4	14.4	18.5	22.5

Appendix 5 California bearing ratio behaviour of OPC+QDbGPC (%) treated soil with 8% CWG

Plunger Penetration (mm)	Plunger Load (kN)								
	California bearing ratio behaviour of OPC+QDbGPC (%) treated soil with 8% CWG								
	0	0+40	5+35	10+30	20+20	25+15	30+10	35+5	40+0
0	0	0	0	0	0	0	0	0	0
0.5	1.4	1.7	2.1	2.3	4.6	8.6	12.6	16.7	20.7
1	1.5	1.8	2.2	2.4	4.7	8.7	12.7	16.8	20.8
1.5	1.7	1.9	2.3	2.5	4.8	8.8	12.8	16.9	20.9
2	1.9	2.1	2.3	2.6	4.9	8.9	12.9	17.0	21.0
2.5	2.1	2.1	2.4	2.7	5.0	9.0	13.0	17.1	21.1
3	2.2	2.3	2.5	2.8	5.1	9.1	13.1	17.2	21.2
3.5	2.3	2.4	2.6	2.9	5.2	9.2	13.2	17.3	21.3
4	2.4	2.5	2.7	3.0	5.3	9.3	13.3	17.4	21.4
4.5	2.5	2.6	2.8	3.1	5.4	9.4	13.5	17.5	21.5
5	2.6	2.7	2.9	3.2	5.5	9.5	13.6	17.6	21.6
5.5	2.8	2.9	3.0	3.3	5.6	9.6	13.7	17.7	21.7
6	2.9	3.0	3.0	3.4	5.7	9.7	13.8	17.8	21.8
6.5	3.1	3.2	3.2	3.5	5.8	9.8	13.9	17.9	21.9
7	3.4	3.5	3.6	3.7	5.9	9.9	14.0	18.0	22.0
7.5	3.8	3.9	4.0	4.1	6.0	10.0	14.1	18.1	22.1
8	4.0	4.1	4.2	4.3	6.4	10.1	14.2	18.2	22.2
8.5	4.3	4.4	4.5	4.6	6.7	10.2	14.3	18.3	22.3
9	4.4	4.6	4.7	4.8	6.9	10.3	14.4	18.4	22.4
9.5	4.6	4.7	4.8	4.9	7.0	10.4	14.5	18.5	22.5
10	4.7	4.8	4.9	5.0	7.1	10.5	14.6	18.6	22.6

Appendix 6 California bearing ratio behaviour of OPC+QDbGPC (%) treated soil with 12% CWG

Plunger Penetration (mm)	Plunger Load (kN) California bearing ratio behaviour of OPC+QDbGPC (%) treated soil with 12% CWG								
	0	0+40	5+35	10+30	20+20	25+15	30+10	35+5	40+0
0	0	0	0	0	0	0	0	0	0
0.5	1.5	1.8	2.2	2.4	4.7	8.7	12.7	16.8	20.8
1	1.6	1.9	2.3	2.5	4.8	8.8	12.8	16.9	20.9
1.5	1.8	2.0	2.4	2.6	4.9	8.9	12.9	17.0	21.0
2	2.0	2.2	2.5	2.7	5.0	9.0	13.0	17.1	21.1
2.5	2.2	2.3	2.6	2.8	5.1	9.1	13.1	17.2	21.2
3	2.3	2.4	2.7	2.9	5.2	9.2	13.2	17.3	21.3
3.5	2.4	2.5	2.8	3.0	5.3	9.3	13.3	17.4	21.4
4	2.5	2.6	2.9	3.1	5.4	9.4	13.4	17.5	21.5
4.5	2.6	2.7	3.0	3.2	5.5	9.5	13.6	17.6	21.6
5	2.7	2.8	3.1	3.3	5.6	9.6	13.7	17.7	21.7
5.5	2.9	3.0	3.2	3.4	5.7	9.7	13.8	17.8	21.8
6	3.0	3.1	3.3	3.5	5.8	9.8	13.9	17.9	21.9
6.5	3.2	3.3	3.4	3.6	5.9	9.9	14.0	18.0	22.0
7	3.5	3.6	3.7	3.8	6.0	10.0	14.1	18.1	22.1
7.5	3.9	4.0	4.1	4.2	6.1	10.1	14.2	18.2	22.2
8	4.1	4.2	4.3	4.4	6.5	10.2	14.3	18.3	22.3
8.5	4.4	4.5	4.6	4.7	6.8	10.3	14.4	18.4	22.4
9	4.5	4.7	4.8	4.9	7.0	10.4	14.5	18.5	22.5
9.5	4.7	4.8	4.9	5.0	7.1	10.5	14.6	18.6	22.6
10	4.8	4.9	5.0	5.1	7.2	10.6	14.7	18.7	22.7

Appendix 7 California bearing ratio behaviour of DOPC+QDbGPC (%) treated soil with 16% CWG

| Plunger Penetration (mm) | Plunger Load (kN) |||||||||
| | California bearing ratio behaviour of DOPC+QDbGPC (%) treated soil with 16% CWG ||||||||
	0	0+40	5+35	10+30	20+20	25+15	30+10	35+5	40+0
0	0	0	0	0	0	0	0	0	0
0.5	1.6	1.9	2.3	2.5	4.8	8.8	12.8	16.9	20.9
1	1.7	2.1	2.4	2.6	4.9	8.9	12.9	17.0	21.0
1.5	1.9	2.2	2.5	2.7	5.0	9.0	13.0	17.1	21.1
2	2.1	2.3	2.6	2.8	5.1	9.1	13.1	17.2	21.2
2.5	2.3	2.4	2.6	2.9	5.2	9.2	13.2	17.3	21.3
3	2.4	2.5	2.8	3.0	5.3	9.3	13.3	17.4	21.4
3.5	2.5	2.6	2.9	3.1	5.4	9.4	13.4	17.5	21.5
4	2.6	2.7	3.0	3.2	5.5	9.5	13.5	17.6	21.6
4.5	2.7	2.8	3.1	3.3	5.6	9.6	13.7	17.7	21.7
5	2.8	2.9	3.2	3.4	5.7	9.7	13.8	17.8	21.8
5.5	3.0	3.1	3.3	3.5	5.8	9.8	13.9	17.9	21.9
6	3.1	3.2	3.4	3.6	5.9	9.9	14.0	18.0	22.0
6.5	3.3	3.4	3.5	3.7	6.0	10.0	14.1	18.1	22.1
7	3.6	3.7	3.8	3.9	6.1	10.1	14.2	18.2	22.2
7.5	4.0	4.1	4.2	4.3	6.2	10.2	14.3	18.3	22.3
8	4.2	4.3	4.4	4.5	6.6	10.3	14.4	18.4	22.4
8.5	4.5	4.6	4.7	4.8	6.9	10.4	14.5	18.5	22.5
9	4.6	4.7	4.9	5.0	7.0	10.5	14.6	18.6	22.6
9.5	4.8	4.9	5.0	5.1	7.2	10.6	14.7	18.7	22.7
10	4.9	5.0	5.1	5.2	7.3	10.7	14.8	18.8	22.8

Appendix 8 California bearing ratio behaviour of OPC+QDbGPC (%) treated soil with 20% CWG

Plunger Penetration (mm)	Plunger Load (kN)								
	California bearing ratio behaviour of OPC+QDbGPC (%) treated soil with 20% CWG								
	0	0+40	5+35	10+30	20+20	25+15	30+10	35+5	40+0
0	0	0	0	0	0	0	0	0	0
0.5	1.7	2.0	2.3	2.6	4.9	8.9	12.9	17.0	21.0
1	1.8	2.2	2.5	2.7	5.0	9.0	13.0	17.1	21.1
1.5	2.0	2.3	2.6	2.8	5.1	9.1	13.1	17.2	21.2
2	2.2	2.4	2.7	2.9	5.2	9.2	13.2	17.3	21.3
2.5	2.4	2.5	2.8	3.0	5.3	9.3	13.3	17.4	21.4
3	2.5	2.6	2.9	3.1	5.4	9.4	13.4	17.5	21.5
3.5	2.6	2.7	3.0	3.2	5.5	9.5	13.5	17.6	21.6
4	2.7	2.8	3.1	3.3	5.6	9.6	13.6	17.7	21.7
4.5	2.8	2.9	3.2	3.4	5.7	9.7	13.8	17.8	21.8
5	2.9	3.0	3.3	3.5	5.8	9.8	13.9	17.9	21.9
5.5	3.1	3.2	3.4	3.6	5.9	9.9	14.0	18.0	22.0
6	3.2	3.3	3.5	3.7	6.0	10.0	14.1	18.1	22.1
6.5	3.4	3.5	3.6	3.8	6.1	10.1	14.2	18.2	22.2
7	3.7	3.8	3.9	4.0	6.2	10.2	14.3	18.3	22.3
7.5	4.1	4.2	4.3	4.4	6.3	10.3	14.4	18.4	22.4
8	4.3	4.4	4.5	4.6	6.7	10.4	14.5	18.5	22.5
8.5	4.6	4.7	4.8	4.9	7.0	10.5	14.6	18.6	22.6
9	4.7	4.8	4.9	5.1	7.1	10.6	14.7	18.7	22.7
9.5	4.9	5.0	5.1	5.2	7.3	10.7	14.8	18.8	22.8
10	5.0	5.1	5.2	5.3	7.4	10.8	14.9	18.9	22.9

Appendix 9 California bearing ratio of OPC+QDbGPC (%) treated soil with CWG

CWG Proportion by wt (%)	CBR of OPC+QDbGPC (%) treated soil with CWG								
	0	0+40	5+35	10+30	20+20	25+15	30+10	35+5	40+0
0	13	14	17	19	36	66	97	128	158
4	14	15	17	20	37	67	97	128	159
8	16	16	18	20	38	68	98	129	159
12	17	17	20	21	39	69	99	130	160
16	17	18	20	22	39	69	100	131	161
20	18	19	21	23	40	70	100	131	162

ABOUT THE AUTHORS

Dr. Kennedy Chibuzor Onyelowe born on December 25, 1978 hails from Umuohia, Ndioulmbe Nvosi in Isiala Ngwa South Local Government Area of Abia State, Nigeria. He acquired a bachelor of engineering degree from the Federal University of Technology, Owerri, Nigeria in 2003. He proceeded to the University of Nigeria, Nsukka where he acquired a master of engineering and doctor of philosophy degrees in Civil Engineering majoring in Geotechnical Engineering in 2010 and 2015 respectively. This was after few years of field practice. He was employed at the Michael Okpara University of Agriculture, Umudike, Nigeria as graduate assistant in 2007 where he progressed through the ranks. For over 10 years, he has burdened himself with research activities, which included investigations into soils re-engineering and computational geotechnics. He leads a research team of experts across the world in his investigations. He has published his findings in reputable journals across the world adding several Thomson Reuters journals to his records of intellectual prowess. His assiduity at what he does has given rise to this book of practice and research, which is dedicated to the cause of research. He is registered with relevant professional bodies, which include Council for the Regulation of Engineering in Nigeria

(COREN), International Society of Soil Mechanics and Geotechnical Engineering (ISSMGE), International Geosynthetic Society (IGS), etc. Dr. Kennedy has a wonderful and adorable wife which has blessed their eternal family with three beautiful children, Favour, Fortune and Fountain.

Engr. Dr. Julian C. Aririguzo has a BEng. degree in Mechanical/Production Engineering and MEng. in Automatic Control and Systems Engineering from University of Sheffield UK. He got a partial scholarship afterwards for a PhD. in Mechanical Engineering specializing in Manufacturing Systems and Engineering.

His research interests and experience revolve around design and analysis of manufacturing systems, next generation of manufacturing systems including Fractal and Biological Manufacturing Systems, computer modeling/simulation of manufacturing systems, and sustainability/green manufacturing. He also regularly works on renewable/green energy projects. He has published several award winning papers in leading international manufacturing conferences and journals based on his research.

In 2008 he won the British council best international students' award (regional category). While teaching and conducting cutting edge research in Eastern Nigeria, Dr. Aririguzo has collaborative relationship and consultancy manufacturing firms and SMEs. He is also the founder and CEO of a non-profit organization.

He is a fellow of the MIT ETT (Cambridge Massachusetts), which affords him ways of bringing his University into better academic collaborative relationship with MIT.

Charles Ezugwu is a Senior Lecturer and Head of Department of Civil Engineering, Federal University, Ndufu-Alike, Ebonyi State, Nigeria. He acquired Bachelor of Engineering degree at former Anambra State University of Technology (ASUTECH), Enugu, Nigeria in 1986. Also, obtained Master of Engineering degree (Water Resources) and Doctor of Philosophy degree (Water Resources), both in Department of Civil Engineering, Nnamdi Azikiwe University, Awka, Nigeria in 2006 and 2013 respectively. He is a member of Nigerian Society of Engineers and registered by the Council for the Regulation of Engineering in Nigeria (COREN).

www.ingramcontent.com/pod-product-compliance
Lightning Source LLC
Chambersburg PA
CBHW030943180526
45163CB00002B/680